World Sustainability Series

Series Editor

Due to its scope and nature, sustainable development is a matter which is very interdisciplinary, and draws from knowledge and inputs from the social sciences and environmental sciences on the one hand, but also from physical sciences and arts on the other. As such, there is a perceived need to foster integrative approaches, whereby the combination of inputs from various fields may contribute to a better understanding of what sustainability is, and means to people. But despite the need for and the relevance of integrative approaches towards sustainable development, there is a paucity of literature which address matters related to sustainability in an integrated way.

Notes on the quality assurance and peer review of this publication

Prior to publication, the works published in this book are initially assessed and reviewed by an in-house editor. If suitable for publication, manuscripts are sent for further review, which includes a combined effort by the editorial board and appointed subject experts, who provide independent peer-review. The feedback obtained in this way was communicated to authors, and with manuscripts checked upon return before finally accepted. The peer-reviewed nature of the books in the "World Sustainability Series" means that contributions to them have, over many years, been officially accepted for tenure and promotion purposes.

Satyanarayana Narra · Khalifa Ababacar Sarr ·
Cheikh Mouhamed Fadel Fall

Sustainable Waste Management in Senegal

Strategies for Collection, Treatment, and Policy Innovation

Springer

Satyanarayana Narra
Material and Energy Valorisation
of Biogenous Residues
Waste and Resource Management
University of Rostock
Rostock, Mecklenburg-Vorpommern
Germany

Khalifa Ababacar Sarr
SONAGED
Dakar, Senegal

Cheikh Mouhamed Fadel Fall
SONAGED
Dakar, Senegal

ISSN 2199-7373 ISSN 2199-7381 (electronic)
World Sustainability Series
ISBN 978-3-031-98967-4 ISBN 978-3-031-98968-1 (eBook)
https://doi.org/10.1007/978-3-031-98968-1

This work was supported by Bundesministerium für Bildung und Forschung.

This Springer imprint is published by the registered company Springer Nature Switzerland AG
The registered company address is: Gewerbestrasse 11, 6330 Cham, Switzerland

Preface

Waste management in most of the developing and emerging countries is associated with various inefficiencies due to the lack of knowledge, available technologies, skilled and trained personnel, governmental policies and enforcement, abundant financial support as well as capacity building. This inhibits the collection, processing and proper valorisation of the secondary products (material and energy). These inefficiencies are stunting the sustainable development in individual countries.

Africa, being one of the biggest continents on earth, is at its pivotal crossroads in the development with nine developed countries (HDI > 0.7), 18 emerging countries (0.5 > HDI < 0.7) and 27 developing countries. Senegal is considered as an emerging country based on the HDI ranking. Senegal's economy and infrastructural development have been continuously growing in the last few years. Eventhough there is a continuous growth, the country still ranks low in the United Nations Human Development Index (HDI) ranking 169 out of 193 countries. This low ranking is due to the food crisis and high unemployment rates in the country. The unemployment rate has indirectly increased the informal sector trying to find alternative means of living, e.g., from waste management. The current practices in the waste management within the country are up to 85% informal. The lack of proper management has resulted in many calamities such as flooding, cholera outbreaks, leakage of wastes into water bodies (lakes, rivers and oceans), etc.

The amount of waste produced is increasing every year due to various factors such as urbanization, increased economic activities and rise in living standards. Based on which waste management is becoming one of the most pressing challenges of the present day in Senegal. Senagal is one of the first African countries to include waste management into their strategic transitional national agenda as released by the Senegalese government in the year 2024 with the title "Senegal 2050: Agenda National De Transformation". The agenda emphasizes the importance of proper and sustainable waste management schemes along with the infrastructure development within the country. The authors have developed a full waste management strategy including various waste streams and their individual treatments as well as their step-by-step implementation enhancing the whole waste management system by integrating all stakeholders in the country.

The book is structured to evaluate the current situation on ground to develop a custom-fit strategy easing the implementation. Subsequent chapters are built based on the specific waste streams to be treated and valorised. The book is structured based on the 8 pillars of the waste management, namely (1) waste minimization, (2) storage, (3) collection gaps, (4) centralization/decentralization, (5) digitalization, (6) capacity building, (7) financial mechanisms, and (8) mitigating emissions. Each chapter brings in innovations in the step-by-step valorization of secondary materials reducing the burden on governmental agencies and enhancing circular economy in the whole country. The book is a close collaborative work with all stakeholders (national agencies, NGOs, Academia, municipalities, collectors, and private and public institutions) involved in the waste management.

Rostock, Germany/Dakar, Senegal Satyanarayana Narra

Acknowledgements The research for this book has been partially funded by the German Federal Ministry of Education and Research under the project "Verbundvorhaben Waste2Energy: Hybrid Energie aus Abfall als nachhaltige Lösung für Ghana".

The authors like to thank their colleagues for their valuable contributions:

Dr. Vicky Shettigondahalli Ekanthalu,
Dr. Qahtan Thabit,
Zipporah Asiedu,
Vidhi Singh,
Prof. Dr. Ahmadu Aly Mbaye,
Dr. Assane Beye,
Dr. Mass Thiam,
Dr. Zeynab Thiam Ndiaye,
Dr. Lhyxzas Tchimboungou.

SPONSORED BY THE

Federal Ministry
of Education
and Research

Competing Interests The authors have no competing interests to declare that are relevant to the content of this manuscript.

Summary

Overview of the Municipal Solid Waste Generation and Management in Senegal

Senegal generates a large amount of municipal solid waste with over 3.1 million tons produced yearly (estimated in 2023). The country is facing a significant obstacle in providing reliable waste management services. Unfortunately 45% of the waste produced (equivalent to 1.3 million tons) remains uncollected. Senegal's average waste collection rate is lower than the sub-Saharan Africa's average collection rate of 65%, where the African average is standing at 45%. It is worth noting that the daily waste generation rate per person in Senegal is consistent with the sub-Saharan Africa's average of 0.65 kg/person per day with an approximation of 0.57 kg. The capital city Dakar (Dakar region) generates the highest annual waste of about 723,108 tons (at a waste generation rate of 0.8 kg per person) followed by Pikine (Dakar region) 181,848 tons and Touba 110,095 tons as recorded in the year 2022. Senegal is estimated to produce 8.95 million TPA of waste by 2050 with a projected national average per capita waste generation increase from 5.0 to 5.7 kg/day and a population growth rate of approximately 3.0% per year. Senegal's waste composition is similar to that of other Sub-Saharan African countries but significantly varies from that of the developed nations. The composition poses a challenge and increases the cost of waste management since it mainly consists of sand, mud and organic debris. The variation in waste composition is notable in different regions ranging from the large urbanized areas such as Dakar to secondary cities. Two-thirds of the waste generated consist of fine particles, sand and organic waste. The remaining one-third includes paper, cardboard, plastics and textiles. The solid waste has considerable potential for energy and resource recovery. The fact that 53% of the waste is non-reusable highlights the importance of sand recycling and sanitary landfills. Waste collection and transportation services in Senegal differ across regions depending on available resources and infrastructure.

Waste collection and transportation services are inadequate in Senegal and limited only to regional capitals. The situation is even worse in areas with high population density and low income where informal collectors charge daily or monthly fees to collect waste using wagons, tricycles or horse carts. Unfortunately access to waste collection is highly uneven with Dakar being the only city with a coverage rate above 50% followed by Rufisque with a coverage rate of only 36%. In regional capitals the collection rate is only 33% and in secondary municipalities it's a paltry 22%. Even when waste is collected in Dakar, it's merely dumped in the Mbeubeuss landfill which is one of the world's ten largest dumpsites. Senegal relies heavily on landfills for waste disposal and in regions without collection services uncontrolled dumping and burning are common. Senegal generates over 2.9 million tonnes of waste annually and over 1.3 million tonnes remain uncollected with the collected waste primarily disposed of at uncontrolled dumpsites across the regions or mainly at the Mbeubeuss disposal site. As Senegal's population growth rate is projected to reach approximately 3.0%, the country could generate an estimated 8.1 million TPA of waste of 38,987,234 inhabitants by 2050 and with no existing waste management system in place there is an urgent need to establish proper waste management practices and expand capacity throughout Senegal. Currently there are no options for collecting sanitary waste from households resulting in high amounts of such waste being either dumped or burnt uncontrollably and hazardous waste generated by hospitals poses serious environmental and health risks yet there are few suitable treatment facilities available for its disposal.

There is a severe shortage of waste storage bins and proper waste collection services in many urban areas due to poor infrastructure and inadequate waste collection vehicles and containers. This has led to escalation in illegal dumping in some municipalities. There are no mechanisms in place throughout the country (except some landfills, e.g., Mbeubeuss) to weigh waste collected from commercial generators, and there are no household or decentralised waste treatment options available. The absence of proper waste management practices in Senegal poses a significant threat to public health and the environment and swift action is needed to address these issues.

In Senegal municipal waste can be classified into distinct pathways for disposal and treatment such as organic waste, recyclable materials, refuse-derived fuel (RDF) potential and non-reusable waste. Effective management of each fraction requires unique approaches. These multi-divisional pathways can assist policymakers and regional stakeholders in making informed decisions reducing stress on landfills and promoting sustainable practices. Currently Senegal generates about 3.1 million TPA of waste as per our calculation (as at 2023 with population growth rate at 3.0%); of this total about 12% (372,000 TPA) of organic waste ends up in illegal dumpsites or landfills and could be treated further. Recyclable waste makes up 25% (775,000 TPA) of the total waste generated with almost half of this fraction (12.5%) being recycled.

Meanwhile RDF potential waste amounts to 387,500 TPA (out of 25 % recyclable waste) and could be used to recover energy but is still being dumped into landfills. Approximately 372,000 TPA of non-reusable waste (12% as inert materials of the total waste composition) also ends up in landfills but the repurposing or recycling the 1,545,578 TPA of waste that is currently dumped could promote sustainable waste practices.

Technical Concept, Strategies, and Approach for MSW Management for Senegal

Senegal's waste management system can be significantly improved by implementing various strategies simultaneously. These strategies have been thoroughly explained offering specific approaches and timelines that can enhance the system's effectiveness. These strategies (Fig. 1) offer designated benefits that can enhance Senegal's waste management system both individually and collaboratively

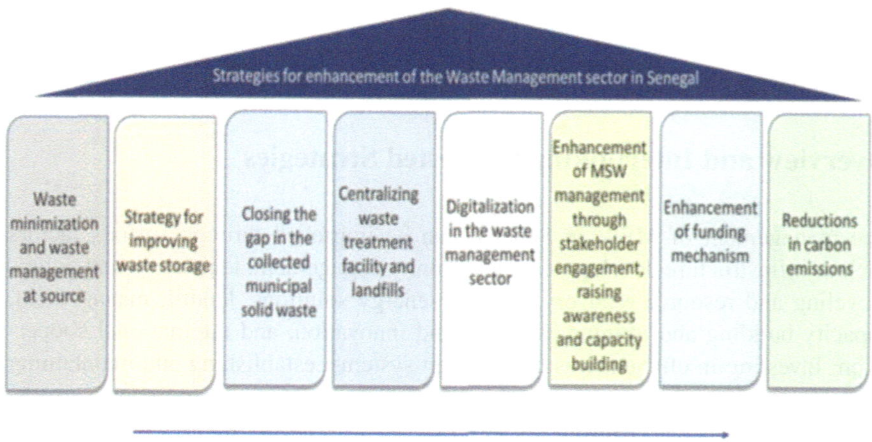

Fig. 1 Technical concept, strategies, and approach for MSW management for Senegal

Presented in Fig. 2 is the linkages and interrelation among the strategies. For instance the use of digital technologies in various stages of waste management including collection, transportation and treatment can offer a holistic approach to managing municipal solid waste (MSW) in a systematic manner. It is advisable to implement all the proposed strategies concurrently to achieve the best possible results.

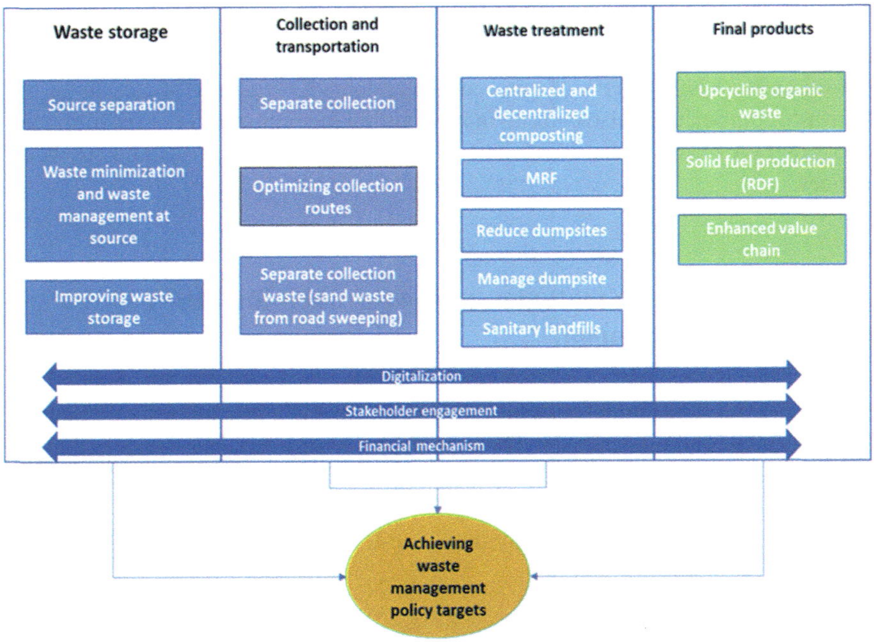

Fig. 2 Overview and interlinking suggested strategies

Overview and Interlinking Suggested Strategies

The management of MSW in Senegal can be improved through various factors such as infrastructure development, community engagement, legislation and policy, recycling and resource recovery, waste-to-energy solutions, landfill management, capacity building and training, research and innovation, and international cooperation. Investing in efficient waste collection systems, establishing and maintaining waste transfer stations, and promoting awareness about waste segregation, recycling, and responsible disposal are key factors. Developing comprehensive waste management policies and regulations, implementing Extended Producer Responsibility programs, and promoting recycling facilities can also contribute to better MSW management. Waste-to-energy solutions involve exploring the potential for electricity generation from organic waste or other combustible materials, investing in modern technologies, and improving landfill infrastructure. Capacity building and training can also be achieved by providing training for waste management personnel and building the capacity of local authorities to plan and implement effective waste management strategies. Supporting research and innovation and collaborating with local universities and research institutions can also help address specific waste management challenges in Senegal. To ensure timely and efficient transport of waste vehicle payloads should be evaluated using a weighbridge or visual estimation. It's crucial to process received waste daily to minimize local disruption and environmental impacts.

Decentralized treatment options can provide additional benefits such as reducing waste dumped at garbage vulnerable points to producing value-added products increasing revenue generation creating more jobs and reducing GHG emissions. Achieving 100% waste segregation throughout Senegal is definitely achievable with the implementation of an optimized collection and transport system. In addition improving collection efficiency offers a plethora of benefits such as reducing the number of garbage-prone areas through waste prevention, increasing the value of waste streams and thereby creating opportunities for Micro, Small and Medium Scale Enterprises (MSMEs) to treat specific waste streams, generating revenue from the sale of valuable products, creating sustainable jobs in waste management and ultimately reducing GHG emissions. Though the interlinking effects may be challenging to quantify the successful implementation of all strategies will undoubtedly yield significant benefits in the short and long terms.

Effective waste management is crucial to combat climate change by reducing greenhouse gas (GHG) emissions. Landfill decomposition is the primary source of GHG emissions from waste management, and it releases methane, a potent greenhouse gas. The strategy strongly recommends implementing diverse strategies to improve waste management and reduce GHG emissions. These strategies can be immediately applied to the current waste management system to enhance waste fraction and its management. However the predicted/projected total wastes by 2050 of at least 8.9 TPA has to be considered and for this three different pathways are proposed. By 2050 a centralized facility should have the capacity to handle at least 5.0 million TPA of waste for material and energy recovery as well as recycling. Any remaining non-reusable waste can be sent to a sanitary landfill that meets the required standards. These landfills can contain up to 3.8 million TPA, including a fraction of the waste from the centralized facilities in Senegal.

Transitioning from traditional landfills (old dumpsites) to sanitary landfills and establishing regional landfills is the way forward to enhance Senegal's waste management system. This strategy offers unique advantages but the success depends on including other waste management strategies that have to be parallelly implemented across the country. Repurposing old dumpsites and creating centralized sanitary landfills can reduce the number of unsanitary and unnecessary dumpsites across Senegal's 14 regions.

Waste management units like recycling and energy recovery plants should be established to reduce the waste footprint and manage final rejects scientifically. A centralized recycling and recovery plant can significantly improve the quality of recycled content and bring substantial environmental benefits to waste management in Senegal. Successful implementation of these strategies will decrease total waste dumped at landfills, offer sustainable alternative fuel sources for industries and organic fertilisers for farmers, create jobs and reduce GHG emissions.

This study has identified numerous significant issues and scientific gaps in waste management. To tackle these challenges it is strongly recommended to implement strategic actions stated in the report. Specifically reducing waste at the source is crucial in alleviating the stress and overload on the waste supply chain and disposal sites. These measures can not only generate revenue and reduce environmental stress but also create employment opportunities. To effectively implement these strategies, it is imperative to have strong governance, policy enforcement and transparent financial flow. By adhering to these recommendations, the waste management sector can be vastly improved leading to enhanced public health, well-being and environmental sustainability in the country. In the end these strategic solutions have the potential to pave their way for a more sustainable and litter-free future for Senegal. It is essential to take action now to ensure that our environment and community remain healthy and thriving.

Contents

Abbreviations

AECID	Spanish Agency for International Development Cooperation
AFD	French Development Agency
BWG	Bulk Waste Generator
EC	Soil conductivity
GIZ	Deutsche Gesellschaft für Internationale Zusammenarbeit
ICT	Information and Communication Technology
m	Meters
MCF	Material Collection Facilities
MSME	Micro, Small, and Medium Enterprises
MSW	Municipal Solid Waste
NGO	Non-Governmental Organisation
OWM	Organic Waste Management
PGDSU	Sustainable Solid Urban Waste Management Project
PROMOGED	Promotion of Integrated Management and Economy of Solid Waste in Senegal
SDG	Sustainable Development Goal
SENELEC	Senegalese National Electricity Company
SOMETA	Société Métallurgique d'Afrique
SONAGED	National Company for Integrated Waste Management
SWM	Solid Waste Management
TPA	Ton per annum
TPD	Ton per day
UCG	Unite de Coordination et de Gestion (Solid Waste Management Coordinating Unit)
UNDP	UN Development Programme
UNFCCC	United Nations Framework Convention on Climate Change
WtE	Waste to Energy

Chapter 1
Introduction

Across the globe waste generation is a rising concern. 2.24 billion of tonnes of solid waste generated in the recent year of 2020, which makes an average of 0.79 kg of waste per person per day. These statistics are predicted to rise at least by 73% by the end of 2050 generating about 3.88 billion tonnes of waste per year. In an outlook developing countries and low-income countries have been reported to have less impactful waste management system where approximately 90% of the waste is disposed openly into illegal dumpsites. This creates an open door for spreading diseases increase in GHG emissions and hazardous accidental damages.[1] Operative waste management is an effective way to deal with these issues. The inhabitants exhibit a lack of awareness regarding the generation of waste and the consequential actions they take that contribute to the expansion of problems associated with municipal solid waste (MSW). This results in mixed-type of waste collection which ultimately ends up being dumped in an informal and unstandardized manner in the landfills. These landfills generate a high amount of GHGs and leachate subsequently polluting atmosphere and the groundwater.[2]

This document is prepared to support the development of the proper implementable and sustainable waste management strategy in Senegal which outlines a comprehensive plan to address the challenges and opportunities related to solid waste management in the west African country of Senegal. Notwithstanding numerous initiatives being taken at the city, state or national level for solid waste management, the issue appears to be still a challenge in the country. These challenges begin with poor waste segregation at the source low coordination between collection and transport points and short of treatment facilities and awareness in people. Solid Waste

[1] Solid Waste Management (n.d.).
[2] Ko et al. (2015).

S. Narra et al., *Sustainable Waste Management in Senegal*, World Sustainability Series,
https://doi.org/10.1007/978-3-031-98968-1_1

Management (SWM) Rules have imposed several obligations on different stakeholders regarding the management of solid waste such as source segregation management of solid waste at household and community levels, transportation of segregated solid waste, setting up of decentralized and centralized solid waste processing facilities and a prohibition on disposal of solid waste in dumpsites/landfills.

This project aims to develop and review the existing state strategy providing overall recommendations and creates a roadmap to embark on the successful implementation of solid waste management sustainably.

1.1 Drawbacks or Challenges in the MSW Management System in Senegal

- The need for immediate waste management options considering the increasing population growth. With a population growth rate of approximately 3.0%, Senegal could generate 8.9 million TPA of waste by 2050. Considering the non-existence of a waste management system in Senegal there is a greater need to start a proper waste management practice and expand the capacity of waste management in Senegal.
- Inadequate waste storage bins, ineffective and scattered waste collection has to be addressed. The lack of waste collection mediums; more primary waste handling vehicles, waste containers and waste collection vehicles are needed at a municipal level throughout Senegal's regional capitals. Many municipalities such as Ngor have no space to position communal containers or assembly points hence the absence of waste ground space is resulting in illegal dumping.
- There is currently no system in place for collecting sanitary waste from households, leading to a significant amount of this waste being dumped or burnt uncontrollably. This is a major concern particularly for hazardous waste generated by hospitals, which can cause severe damage to both the environment and human health. The licensed businesses in Dakar do collect hazardous wastes, but the final destination of this waste is often unknown due to either a lack of suitable treatment facilities or obsolete facilities or proper tracking system.
- There are very few or no decentralised or centralised municipal solid waste (MSW) treatment facilities such as composting, Material Recovery Facility (MRF), Refuse Derived Fuel (RDF) and waste-to-energy plants across the country. It is important to consider waste especially MSW as a source of energy and fuel that can be used in different ways and with different technologies. The concept of waste-to-energy can decentralise energy generation for rural areas by converting the waste generated daily by inhabitants into useful energy.
- Settlements create small and uncontrolled dumpsites due to improper waste collection systems and no availability of landfills. The closure and reclamation of these dumpsites can be done with greater ease that could not only result in the reduction

of waste footprint but also has the potential to make more land space available for SWM and other additional activities.

1.2 Objective

This study seeks to strengthen the entire capacity of solid waste management in Senegal. The detailed strategies presented in this report is developed after the comprehensive desk research (review of the current policies and laws as well as existing technologies and management options and programs related to solid waste management).

The developed strategies include the recommendations to guide Senegal's solid waste management system such as:

- Strategies for the utilization of appropriate solid waste treatment technologies and types of facilities including recommendation on siting and replication of the facilities.
- Strategies for solid waste diversion. Recommendations for separate collection, transportation of solid waste and recommendations on collection fleet etc.
- Recommendation for the transition from dumpsite to sanitary landfills and suggestions on hotspot creation
- Extended approaches for the development of solid waste management system in Senegal through the developed strategies

The developed strategies foster the countries efforts for managing solid waste sustainably. The successful implementation of the developed strategies has greater potential for improving the overall solid waste management infrastructure in Senegal.

References

Ko JH, Xu Q, Jang YC (2015) Emissions and control of hydrogen sulfide at landfills: a review. Crit Rev Environ Sci Technol 45(19):2043–2083

Solid Waste Management (n.d.) Retrieved 17 May 2023, from https://www.worldbank.org/en/topic/urbandevelopment/brief/solid-waste-management

Chapter 2
General Information on Waste Management Sector of Senegal

The Republic of Senegal is a country situated on the west-end tip of the African continent. Senegal is also known as 'Gateway to Africa'. At present the area of Senegal was once a part of the country Ghana it was a gate way to sub-Saharan regions of Africa. It was the nodal point to connect Africa with England, France and Portugal.[1] It is spread over an area of 196,772 km^2.

It shares the border with the countries of Mauritania in the north, Mali in the east, Guinea and Guinea-Bissau in the south-western region, Gambia within the country interiors on the southwest. The rest of the border area is a coastal line with Atlantic Ocean.[2] Senegal became an independent country in 1960 and has been marking peace and stability with respect to other African countries.[3] Senegal is divided into 14 regions which are Dakar, Diourbel, Fatick, Kaffrine, Kaolack, Kédougou, Kolda, Louga, Matam, Saint-Louis, Sédhiou, Tambacounda, Thiès and Ziguinchor. Figure 2.1 displays the map of Senegal. The capital city of the country is Dakar.

According to UN Development Programme's (UNDP) in 2006 Senegal was ranked 156th out 177 countries across the globe. It was mainly ranked because of the dominant poor rural regions of the country where almost 60% of populations resides. Besides poverty climate change is also an alarming agenda for the socio-economic structure of the country. The prosperity index of Senegal has made a lot of progress in all sectors in the recent years and currently ranks 102 out of 167.[4]

---

1 Senegal | Culture, History, and People | Britannica (n.d.).

2 Key to Map Symbols (n.d.).

3 Senegal Overview: Development news, research, data | World Bank (n.d.).

4 The Legatum Prosperity Index in 2023, https://www.prosperity.com/globe/senegal (last visited on 1.7.23).

S. Narra et al., *Sustainable Waste Management in Senegal*, World Sustainability Series,
https://doi.org/10.1007/978-3-031-98968-1_2

Fig. 2.1 Senegal map—cities and roads[5]

2.1 Population/Demography

As of April 2019, Senegal has 17.2 million habitants with Fig. 2.3 demonstrating the distribution of Senegal's population density across the country. This number contributes to 0.21% of the total population of the world and it ranks 70th in the world by its population number. The population density is 87 person per Km^2 (225 per mi^2). Out of the total population almost 47.2% of the people live in the urban areas and the rest of them in the local/rural areas[6] (see Fig. 2.2). It is one of the most industrious countries in the west African regions. The country generates its major revenue from agriculture, industry and other devoted services with the share of 15.5, 21.7 and 63.3% respectively.

In comparison to other countries of Africa, Senegal lacks stable supply of natural resources from its own land. Its highest revenue is generated from fishing, tourism and services.[7]

---

[5] Source: GISGeography.com.

[6] Senegal Population (2023).

[7] Solid Waste Management City Profile Climate and Clean Air Coalition Municipal Solid Waste Initiative (n.d.).

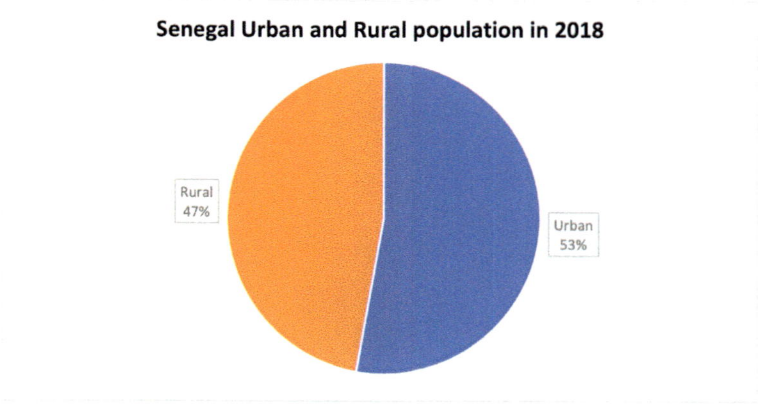

Fig. 2.2 Urban and rural population of Senegal

2.2 Geographical Condition

As Senegal falls in the sub-Saharan region, it possesses diverse climatic conditions. In the north it has desert areas and more tropical in the south. The seasons are typically divided into two patterns each followed by the other. One is a dry season which begins in November and ends by May. The second season is the humid season which begins in June and ends in October. The humid season is also called as wintering with higher temperatures mixed with rains. The coastal areas are the coldest and the most eastern parts are the hottest of all.[8]

Senegal is a predominantly flat country located in the Senegal-Mauritanian Basin. The terrain is characterized by three main structural divisions. The Cape Verde headland situated in the western part of the country consists of small plateaus made of volcanic rock. In the southeast and eastern regions there are fringes of ancient massifs which are mountain masses connected to the Fouta Djallon massif on the Guinea border. The highest point in Senegal reaching an elevation of 1906 feet (581 m) near Népen Diakha is found in this area. The remaining central region is a large shallow landmass lying between Cape Verde and the eastern edges of the massif. The Atlantic coast of Senegal washed by the Canary is currently sandy and subject to strong surf. The coastal south of the Cape Verde Peninsula experiences less intense waves while the coastline south of the Saloum River consists of rias (drowned valleys) and is increasingly bordered by mangroves.[9]

[8] Geography of Senegal General information and National Parks (n.d.).

[9] Senegal–Land | Britannica (n.d.).

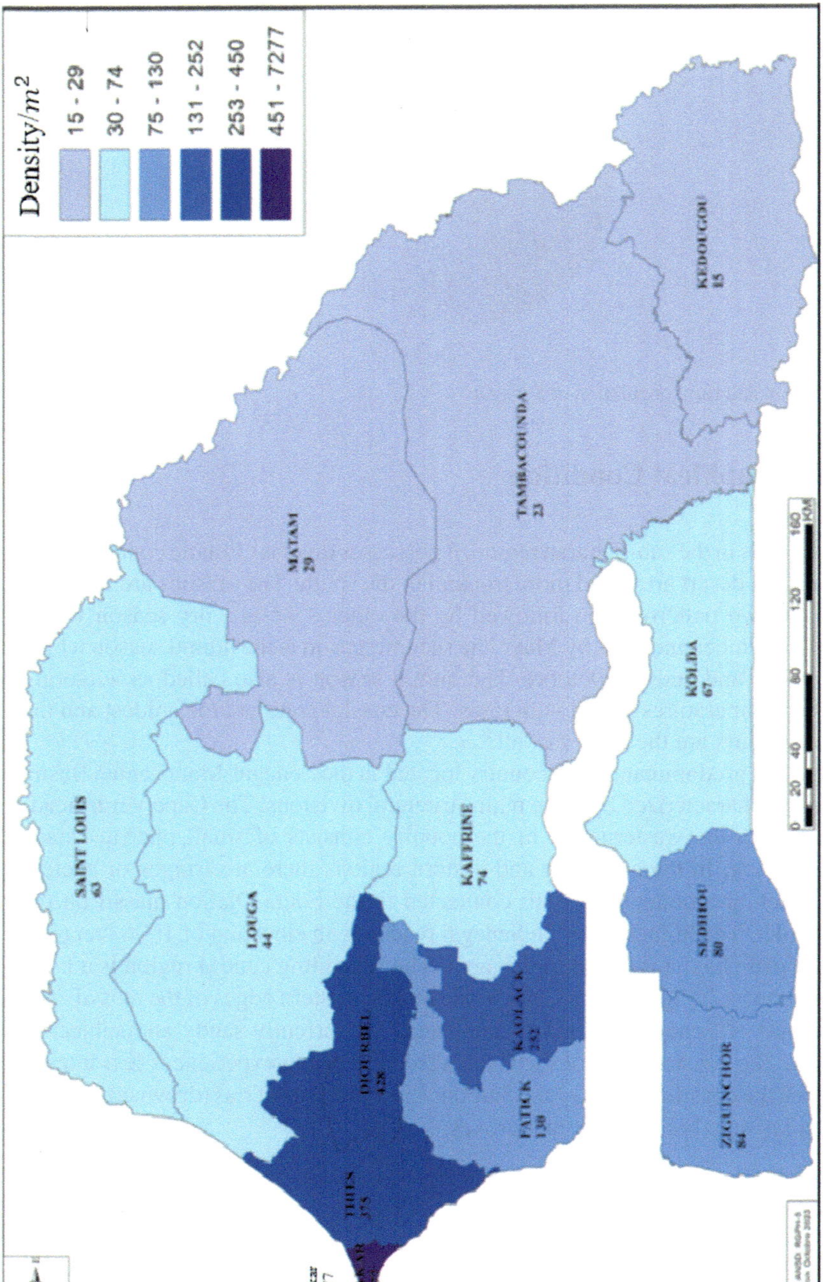

Fig. 2.3 Population density map of Senegal (this figure was adapted from "report preliminary 2023" National Agency of statistics and Demography)

Fig. 2.4 Climatic zones of the Senegal[10]

Senegal experiences areas with tropical climate (depicted in blue, Fig. 2.4), arid (steppe) climate as well as arid (desert) climates.

Soil and Soil Quality: Senegal exhibits significant soil diversity despite its apparent uniformity. The country's soils can generally be classified into two main types: valley soils and soils found in other regions. The alluvial soils in the middle courses of the Sénégal and Saloum rivers are predominantly sandy loams or clays. Near the river mouths these soils tend to be saline and are well-suited for grazing. Similar conditions can be observed along the Gambia and Casamance rivers where the banks are muddy near the mouths and the upper courses have sandy clay soils.

Senegal's northwest region is characterized by ochre-coloured light soils known as Dior soils. These sandy soils combined with iron oxide are highly favourable for peanut cultivation. The dunes formed by these soils provide excellent conditions for growing peanuts while the soils between the dunes are suitable for other food crops such as sorghum. The plateau in the south west consists of sandy clay soils that often undergo laterization resulting in the development of red iron-bearing soils. In the central and southern parts of the country a layer of laterite is concealed beneath a thin sand cover offering limited grazing opportunities during the rainy season. In the Casamance area heavily leached clay soils with a high iron-oxide content are predominant and are suitable for cultivation regardless of their depth.

[10] https://www.bluegreenatlas.com/climate/senegal_climate.html.

2.3 Current Policies and Initiatives

PNGD (National Waste Management Programme)[11]: Government of Senegal in 2013 initiated National Waste Management Programme as an assistance to the municipalities handling the local waste in their regions. The focus of the programme is two directional; one is by providing investment to the local municipal body and the other is by backing up the local government in the institutional, technical and financial capacity. It is driven to achieve at least 75% accessibility of the sanitation services to the citizens as a public health concern. It also objectifies to create revenue by recycling and reducing waste to over EUR 68.43 million (XOF 45 billion) and also to create 15,000 jobs throughout the country.

SONAGED—National Company for Integrated Waste Management[12]: With the unanimous voting SONAGED was created in June, 2022 under the Ministry of Town Planning, Housing and Public Hygiene to replace **Solid Waste Management Coordinating Unit (UCG)**: SONAGED aims to promote inter municipal cooperation, public–private partnerships and the circular economy for modern waste recovery infrastructure. This establishment will professionalize the sector combating youth unemployment and facilitate decentralized waste management under the programme "zero waste Senegal" by strengthening waste management services, creating sustainable cleanliness management and developing a waste recovery industry. SONAGED is fully owned by the state and is expected to create over 17,000 jobs and treat 2.5 million tons of waste annually. An initial investment of 200 billion FCFA will support the company's operations and services across 557 local authorities.

Waste to Energy Initiative and deployment[13]: The government of Senegal has recently launched the Zero Waste Programme in the pursuit of universal electrification by the year 2025. This programme has given a push for the first ever waste-to-energy large scale plant in Senegal. The facility is operated by the Senegalese National Electricity Company (SENELEC) in partnership with the newly formed National Integrated Waste Management Company (SONAGED). With a population consistently growing over 2.5% annually Senegal faces a rising issue of plastic waste generating over 200,000 tons per year of which only 9,000 tons are recycled. The waste-to-power plant in Kaolack aims to address this by recycling the annual plastic waste generation per capita through a thermal power station helping close the country's power demand–supply gap. The project is part of Senegal's broader sustainable development initiatives transforming Kaolack into a beacon of streamlined sustainable development.

---

[11] Sector Brief Senegal: Solid Waste Management and Recycling. GIZ, 2021.

[12] https://www.urbanisme.gouv.sn/actualites/gestion-des-ordures-sonaged-sa-prend-le-relai-de-l%E2%80%99ucg.

[13] Waste to Energy Program Brings Fresh Power Boon to Senegal (n.d.).

PROMOGED (Promotion of Integrated Management and Economy of Solid Waste in Senegal)[14]: PROMOGED was officially launched on June 24, 2021 at Mbuebeuss, a large landfill in the outskirts of Dakar capital city. This project aims to enhance the governance of solid waste management in selected cities. It focuses on domestic waste and its supply chain by promoting inter-municipal cooperation. It also focuses on using waste to create a socio-economic balance by generating revenue and capacity building, thus extending a hand toareas the private sector. This program is co-financed by World Bank, the French Development Agency (AFD) and Spanish Agency for International Development Cooperation (AECID). In 2020, the provisional budget was XOF 170 Billion (EUR 259.16 million). The project is initially limited to six years.

Priority Cleanliness Program (3P—*Le Programme Prioritaire de Propreté*): The Priority Cleanliness Program (3P) was initiated in 2014 to address widespread insalubrity and illegal dumping. It involved 53 local authorities and successfully eradicated illegal dumping, cleared roadways and improved waste collection. The program improved public health by increasing access to cleaning services, achieving higher collection and sweeping rates and disposing of over 1 million m3 of waste in landfills. A Geographic Information System was implemented to map illegal dumping and infrastructure. Senegal has more than 40 uncontrolled dump sites that are harmful to the environment (see footnote 11). Technical committees were established to coordinate waste management at the communal level. Agreements and plans were developed to guide the program's objectives.

References

À propos du PROMOGED | PROMOGED—Projet de Promotion de la Gestion intégrée et de l'Économie des Déchets Solides au Sénégal (n.d.). Retrieved May 7, 2023, from https://promoged.sn/fr/propos-du-promoged

Geography of Senegal General information and National Parks (n.d.) Retrieved 19 Apr 2023, from https://www.bouelmogdad.com/geography/

Key to Map Symbols (n.d.) Retrieved 19 Apr 2023, from www.worldclim.org/current

Senegal–Land | Britannica (n.d.) Retrieved 8 May 2023, from https://www.britannica.com/place/Senegal/Land

Senegal | Culture, History, and People | Britannica (n.d.) Retrieved 19 Apr 2023, from https://www.britannica.com/place/Senegal

Senegal Overview: Development news, research, data | World Bank (n.d.) Retrieved 19 Apr 2023, from https://www.worldbank.org/en/country/senegal/overview#1

Senegal Population (2023) Worldometer (n.d.) Retrieved April 19, 2023, from https://www.worldometers.info/world-population/senegal-population/

Solid Waste Management City Profile Climate and Clean Air Coalition Municipal Solid Waste Initiative (n.d.) Retrieved 7 May 2023, from www.decentralisation.gouv.sn

Waste to Energy Program Brings Fresh Power Boon to Senegal (n.d.) Retrieved 5 May 2023, from https://energycapitalpower.com/waster-to-energy-power-boon-senegal/

---

[14] À propos du PROMOGED | PROMOGED—Projet de Promotion de la Gestion intégrée et de l'Économie des Déchets Solides au Sénégal (n.d.).

Chapter 3
Status Quo on the Organizational Structure of Waste Management

3.1 Overview of the Stakeholders Involved in Waste Management

Throughout Senegal the organisational structure as observed was staged into three different levels of stakeholders. It begins with national ministries which are responsible for laying and managing national policies, laws and strategies. On the second level are the regional authorities which are spread over 14 main regions of Senegal for their administrative occurrence. The third category is where the local, private and non-governmental sector plays a role. It is a combination of municipal authority which is the formal sector and the private organisation as an informal sector.

3.1.1 National Stakeholders

Ministry of Environment and Sustainable Development: In Senegal, the Ministry of Environment and Sustainable Development is responsible for guidance on waste management.[1] The legislation known as the Single Use Plastics Prohibition Law (No. 2020–04) was enacted by the government of Senegal in 2020. The utilization of single-use plastic items, including cups, lids, and pipettes, is strictly forbidden by law.[2] The Ministry of Environment and Sustainable Development is also responsible for laying out the policies and their implementation for the protection of nature, flora and fauna. It also fights against the pollution and keeps an environmental watch. It has a direct authority over national parks and the protected areas. It connects and gives power to local communities for the protection of flora and fauna on and off the land.

[1] Kaza et al. (2018).

[2] WACP program, 2019.

S. Narra et al., *Sustainable Waste Management in Senegal*, World Sustainability Series,
https://doi.org/10.1007/978-3-031-98968-1_3

Ministry of the Economy, Planning and Cooperation: In the year 2021 European Investment Bank (EIB) formally agreed with the Ministry of Economy Planning and Cooperation to invest CFAF 75 Billion (Euro 114.6 million) for the improvement of water supply and integrated solid waste management system in three cities of Senegal (Saint-Louis, Kaolack and Kolda).

3.1.2 Regional Stakeholders

SONAGED—National Company for Integrated Waste Management[3] (Previously "Unite de Coordination et de Gestion (UCG) and currently SONAGED): It is mainly responsible for waste collection and transportation services in the regional and departmental capitals. Usually municipal units provide them utilities for management of the waste. Their roles are as follows:

- Coordinate, monitor and implement regulations and policies mentioned.
- Develop and implement projects for the betterment of the existing SWM system.
- Mobilize and allocate the sector's financial resources.
- Ensure the piloting of the delegated project management or the assistance to management of solid waste projects and programs.
- Ensure capacity building of territorial actors in the field of solid waste management.
- Develop and implement communication, awareness and social mobilization plans.
- Promote the involvement of the private sector in the waste management system.
- Promote solidarity and cooperation between local authorities by communities through consultation frameworks bringing together all the stakeholders of the waste management involved.

3.1.3 Local/Private/NGO Stakeholders

Islamic development bank[4]: In January 2014, Islamic development bank launched Sustainable Solid Urban Waste Management Project (PGDSU) for the enhancement of solid waste management system in the capital city of Dakar with the regional focus on Tivaouane, Kaolack and Touba Mosquée. The project is supported by a funding of EUR 26.67 million (XOF 17.5 billion).

[3] National Strategy for Integrated solid waste management in Senegal (2018–2026) https://www.waste-ndc.pro/wp-content/uploads/2021/06/Strate%CC%81gie-de-gestion-inte%CC%81gre%CC%81e-et-durable-des-de%CC%81chets-UCG-2106-1.pdf.

[4] https://global-recycling.info/archives/6565.

3.2 Legal Framework, Responsibilities, and Organizational Structures

Environment Code (Law 2001–01)[5]: The law came into existence from 12th April 2001. It states clear rights for the citizens of Senegal to have clean and healthy environment at the same time reminds their responsibility for its protection. It generally mentions the importance of biodiversity protection of forests and its management in air pollution planning of the urban areas and disposal of hazardous wastes all together.

Plastic Law—The prevention and reduction of impact on the environment from Plastic Products[6]: The plastic law was initially adopted on 4th May, 2015 by the Ministry of Environment and Sustainable Development and later the law was amended with further inclusions in the year 2020 (Table 3.1). The law applies to all the products made up of plastic whether or not intended for single use. The law is imposed in order to ban single-use plastics such as beverage containers cutlery plates straws and water carrying sachets. It also initiates a plastic recycling-deposit and refund system for the plastic bottles. Under this law all the plastic producers must bear their name on the plastic product and all producers should responsibly collaborate to reduce its generation and recycle by collecting from its end user sources.

National Strategy for Integrated and Sustainable Management of Solid Waste (2018–2026): The strategy is the improved version of the Waste Management Strategy for the years 2013 to 2018 providing clear objectives strategic options and implementation benchmarks for achieving sustainable waste management in Senegal. The Senegalese government specifically the State Coordination Unit (SCU) has developed a strategy for integrated and sustainable solid waste management. This strategy aligns with the national development frameworks including the Emerging Senegal Plan (PSE) and Act III of decentralization which emphasize building viable and sustainable territories through effective implementation of public policies.

Strategies for reducing GHG emission: Reducing greenhouse gas (GHG) emissions through improving waste management systems is an effective strategy to combat climate change. Waste management contributes to GHG emissions primarily through the decomposition of organic waste in landfills and the release of methane a potential greenhouse gas.

---

[5] Nachmany et al. (n.d.).

[6] Senegal: Select plastic laws | ELAW (n.d.).

Table 3.1 Latest inclusion and update on the Plastic law from 2020 relating to the prevention and the reduction of the environmental impact of plastic products (Senegal: Select plastic laws | ELAW, n.d.)

<table>
<tr><td>Restriction on the selling of plastic and plastic products</td><td colspan="2">• It includes a restriction on the production import sale offering a sale use in any form single use or not and even disposable plastic
• It does not include a ban on plastic for food packaging handling and transport of materials from vendor to consumer</td></tr>
<tr><td>Setting the point system</td><td colspan="2">• A deposit should be made in order to purchase any product contained in plastic bottles. The seller is required to bring the plastic carrier of the product back to a collection point in case of the return of the deposit made initially
• The producer should report in every 6 months to the Minister in charge for the environmental sector. This report should state the quantity of plastic bottles places and replaced in market and number of collection points set-up and active</td></tr>
<tr><td>Prevention of production of selective waste and recycling</td><td colspan="2">• Producers are obliged to reduce waste at the source if not, they are likely to be recycled or recovered in conditions it has no impact on the environment
• Producers are obliged to include the recycled plastic waste to the new products of the plastic produced. For these national targets should be set up
• Consumers after the consumption of the product are responsible to sending the plastic containers/plastic products to the collection points</td></tr>
<tr><td>Financial Provisions in the tax</td><td colspan="2">• A fixed price is set up for the plastic recycling companies to buy kilogram of plastic waste for recycling
• 'Plastic tax' is levied on the products made up on non-recycled plastic</td></tr>
<tr><td rowspan="4">Penal dispersions</td><td>Manufacturing Plastic</td><td>Anyone of who manufactures or imports single-use plastics shall be punished by an imprisonment of 11–12 months and/or fine of 5–7 million CFA francs</td></tr>
<tr><td>Consuming plastic products</td><td>At the consumer end, if anyone uses single-use plastic product shall be punished 1–3 months imprisonment and/or 50,000 to 100,000 CFA francs</td></tr>
<tr><td>Not setting up enough collection points</td><td>3–6 months of imprisonment and/or 5–10 million CFA francs</td></tr>
<tr><td>Dumping plastic waste illegally</td><td>Any producer or consumer if illegally dumps the plastic waste at any other place than the collection point is viable for a punishment of 15 days to a month of imprisonment and a fine of 20–50 million CFA francs</td></tr>
<tr><td colspan="3">The formulation of a plan for the implementation of alternatives has not yet occurred. As a result of Covid-19, the anticipated restriction on single-use water bags has been delayed in order to reduce the danger of transmission from shared water containers. Government initiatives concerning single-use plastic will necessitate a significant emphasis on public awareness in order to get public support. Present awareness mostly pertains to the consequences of littering, neglecting to encompass the advantages that can be gained from adopting a circular economy[a]</td></tr>
</table>

[a] Circular Economy in Africa-EU Cooperation—Country Report Senegal. 2020

References

Kaza S, Yao L, Bhada-Tata P, Van Woerden F (2018) What a waste 2.0: a global snapshot of solid waste management to 2050. Urban Development Series. World Bank, Washington, DC

Nachmany M, Fankhauser S, Davidová J, Kingsmill N, Landesman T, Roppongi H, Schleifer P, Setzer J, Sharman A, Singleton CS, Sundaresan J, Singleton TTS (n.d.) Climate change legislation in Senegal. The 2015 Global climate legislation study a review of climate change legislation in 99 countries. Retrieved 7 May 2023, from www.lse.ac.uk/GranthamInstitute/legislation/

Senegal: Select plastic laws | ELAW (n.d.) Retrieved 5 May 2023, from https://elaw.org/plastic/SN_PlasticLaws

Chapter 4
Technical Concept, Strategies and Approach for MSW Management

4.1 Structural Overview of Strategies and Approach for MSW Management

Senegal's waste management system can be greatly improved by implementing multiple strategies at the same time. These strategies have been thoroughly explained and offer specific approaches and timelines that can enhance the effectiveness of the system. By adopting these strategies, Senegal can enjoy a wide range of benefits that would improve the waste management system both individually and collaboratively.

Strategic actions are suggested based on these major issues and some scientific identified gaps found during this study. The various strategies to be followed and implemented are shown in Fig. 4.1. The strategies are explained in the all sub-sections of the various chapters in this report. All the explained strategies have their specific designated benefits to enhance the waste management system and the implementation of these strategies will also have an individual and collaborative impact on the overall management of municipal waste in Senegal in the broader context. The interdependence of the strategies and the connection between the strategies, for instance the implementation of digital technologies through the different phases of waste management such as waste collection, transportation and treatment systems provides a holistic approach to the systematic management of MSW as shown in Fig. 2.2. To this extent when possible all the suggested strategies have to be initiated simultaneously to achieve the greater and effective outcome. In order to enhance the situation of MSW treatment in Senegal, it is imperative that the SONAGED and MSW management authorities prioritize waste minimization and treatment at the source. Comprehensive material recovery should occur prior to the final proper disposal of the generated waste. SONAGED must promote and implement activities such as animal feeding, home composting, recycling units and plan decentralized and centralized facilities. This approach would help decrease the burden on the MSW management system.

S. Narra et al., *Sustainable Waste Management in Senegal*, World Sustainability Series,
https://doi.org/10.1007/978-3-031-98968-1_4

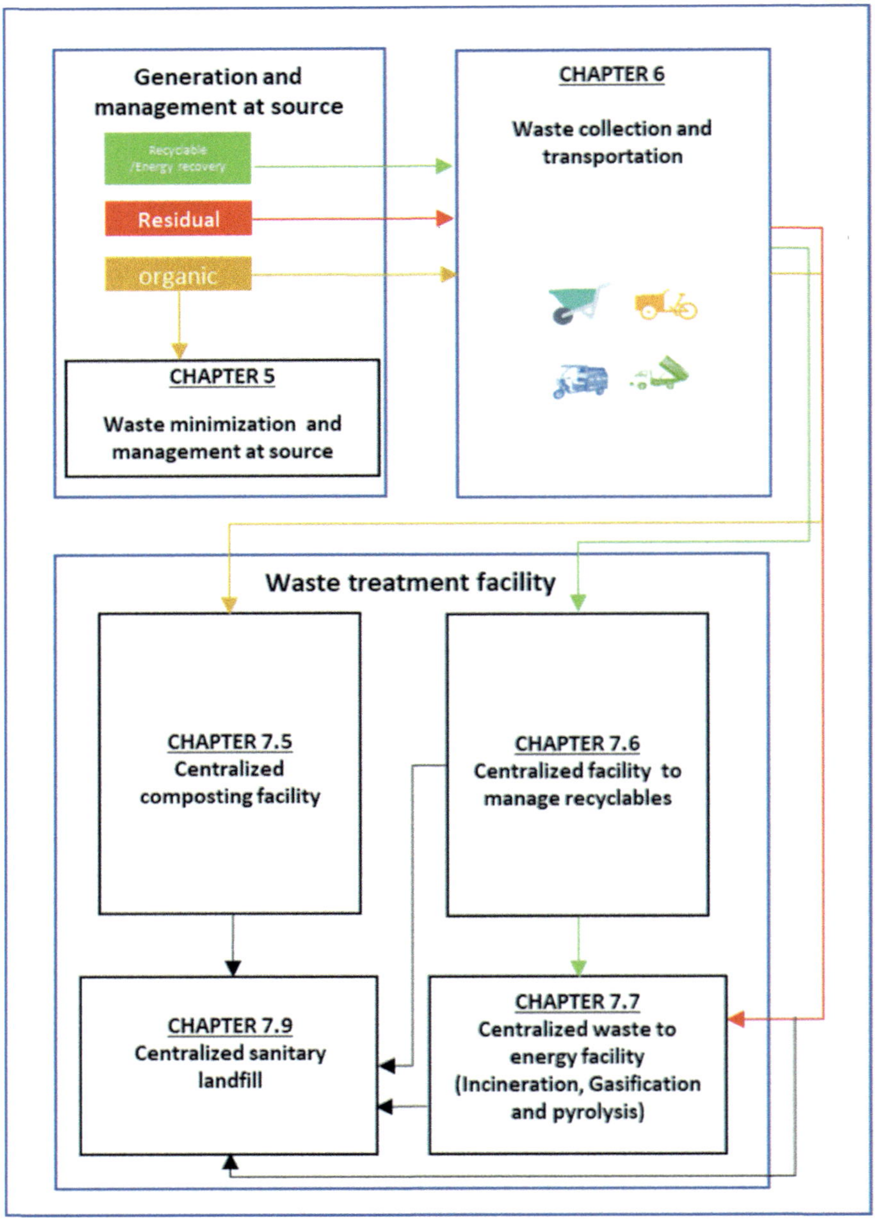

Fig. 4.1 Sections of the proposed strategies for the enhancement of the waste management sector in Senegal

The first strategy which is "The strategy for waste minimisation and treatment at source" can be implemented by source segregation of waste and having clear waste streams collected separately. Waste prevention or minimisation implicates actions taken before a substance material or product becomes waste. The concept of waste minimization is ideally to prevent waste generation by management of waste surplus at all levels of the value chain from the farm (producer) to the consumer. Waste prevention actions result in the reductions of waste quantity generated. While having the clear waste streams, the MSW stream should be focused to recycle the material and nutrients. To this extent involvement of circular economy principles are important in MSW sector. The use of circular economy principles in municipal solid waste management encompasses the following objectives: reduction of waste production, encouragement of material recycling and reuse, energy recovery from waste, and appropriate disposal. The objective of this methodology is to establish a sustainable cycle in which materials are perpetually reintroduced into the manufacturing process, waste is minimized, and resources are utilized efficiently; thus, environmental harm is decreased and long-term resource conservation is promoted.

To initiate a concrete base for the futuristic application of policies, a database including all levels of information should be created and updated at a regular interval. Being aware of what exactly is being discarded, there is a higher chance to reduce disposal costs, reduce over-purchasing, reduce labour costs, reduce the amount of water and energy used for production and reduce greenhouse gas emissions. For instance concerning the organic waste, home composting is an easy way to manage organic waste at home and contribute to waste reduction efforts. This method helps divert organic waste from landfills reducing the cost of solid waste management. After reducing the avoidable sources of MSW waste a second focus has to be given to enhancing the reuse mechanism of generated organic waste. Implementation of extended producer responsibility could make the producer obligated overseeing the complete life cycle of a product. EPR provides financial incentives to producers, encourages product design and packaging optimization, and promotes recycling and waste collection infrastructure improvement. Furthermore, it fosters consumer education and consciousness regarding the ecological consequences of their decisions, supports inventive approaches to prolonging the life of products, and advocates closed-loop systems that involve the perpetual recycling and reuse of materials. EPR promotes a more sustainable paradigm in both production and consumption by holding producers responsible for the environmental consequences of their goods, thereby leading to a reduction in waste generation.

It's important to process the receiving of waste daily to minimize local disruption and environmental impacts. Beyond that there are further benefits to be gained through the successful deployment of a decentralised treatment options to treat the waste fraction of MSW generated (1) decrease the waste dumped at garbage vulnerable points (2) value added products are produced, (3) increase revenue generation (4) creation of more jobs (5) reduction in GHG emissions etc. In the same way 100% waste segregation throughout Senegal could be achieved by means of an optimised collection and transport system but before this, the storage of wastes in the right separated bins are needed. The expected categories of waste at the household level are organic, dry and hazardous domestic waste. This waste can be easily stored in separated bins with their allocated labels of coloured bins. Bins placed for the cumulative collection of waste at the common point vendors should be compatible in size, capacity and weight with the primary collection vehicle.

The selection of a proper waste vehicle type and waste collector teams is another important aspect of collection service where a strategy has been suggested for SONAGED. The type of collection vehicle is dependent on various aspects which include population density, road size, financial liability and waste volume. Participation of the community is crucial for a successful implementation of waste management strategies. The citizens should be contacted through several channels such as community leaders, face-to-face interaction, awareness campaigns, advertising tools and integration of awareness-raising on complete waste management. Chapter 9 elaborates on the strategy for awareness-raising. Achieving 100% collection efficiency will benefit from a number of advantages. (1) a significant decrease in the number of garbage vulnerable points through waste prevention (2) increases the value of waste streams (reuse recycling and energy recovering of waste streams) enable the creation of MSMEs to treat the specific waste stream (3) increase the revenue from sale of value-added products from waste (4) creation of more sustainable jobs in every aspect of the movement of the waste in its management chain (5) reduction of GHG emissions etc.

The interlinking effects could be hard to quantify and the successful implementation of all the strategies will have a significantly increased interlinking benefit that could be seen in a shorter and longer period. To improve Senegal's waste management system a transition from dumpsites into sanitary landfills and suggestions for the setting up of Regional Landfills have been given particular special benefits. The implementation of such a strategy will have an interlinkage effect with other strategies that allow MSW management throughout the country to be carried out on a wider scale. For instance, as mentioned in Sect. 7.9 a direct effect of creating a centralized hotspot of sanitary landfills and reclaiming the old dumpsites can be seen through a decrease in the number of unsanitary and unnecessary dumpsites across the 14 regions in Senegal. The potential to reduce the waste footprint and scientific management of the end of life products of the final rejects produced in the country is by making more land available for the establishment of waste management units, such as a recycling and energy recovery unit. The implementation of a centralized recycling and recovery facilities as set out in Sects. 7.5, 7.6 and 7.7 shall contribute to improving the quality of recycled content along with achieving optimum environmental benefits for waste management practices in Senegal. The successful implementation of a central sanitation landfill combined with recycling and energy recovery facilities will also give rise to other benefits such as (1) decrease the organic waste dumped at garbage vulnerable points, (2) increase the quality of compost, (3) provide sustainable alternative fuel source for industries, (4) socio-economic benefits, (5) creation of more jobs, (6) reduction in GHG emissions etc.

Digital technologies are crucial in an effective waste management sector for a safer, transparent, economically feasible and highly resource-efficient waste management system. Chapter 8 suggest a strategy and timeline for digitalizing the waste management sector. Various digital technologies in the waste management sector have been introduced and discussed. The main advantages of digitalization in reducing paper works have cost reduction more storage of data easiness to find data and reduction in chances for errors.

Generally the interlinking effects could be hard to quantify and the successful implementation of all the strategies will have a significantly increased interlinking benefit that could be seen in a shorter and longer period.

The proposed waste management strategies in Senegal can be implemented in phases, starting with the capital city of Dakar and gradually extending to other regions. This controlled and monitored approach ensures that lessons learned from Dakar can be applied to the entire country. The first phase involves pilot implementation in Dakar, introducing waste minimization and treatment strategies, decentralized treatment options, optimized collection and transport systems, community participation, recycling and recovery facilities, and digital technologies for waste management. The second phase involves evaluation and refinement, focusing on the outcomes and challenges of the strategies implemented in Dakar. Once refined, the strategies are gradually extended to other regions, tailoring them to the specific characteristics and needs of each region. Community participation and awareness programs are also implemented, and the transition from dumpsites to sanitary landfills is continued. Digital technologies are expanded throughout Senegal. Continuous monitoring and improvement are essential to ensure the effectiveness and sustainability of the waste management strategies. This approach maximizes the chances of success, minimizes risks, and creates a model for sustainable waste management practices nationwide.

4.2 Waste Generation and Composition in Senegal

The lack of accessibility to services for the disposal of solid waste is a persistent problem in Senegal. Senegal generates more than 2.9 million tons of solid waste every year. One of the most persistent challenges facing the country is that it does not have access to dependable waste management services. Out of the total generated waste approximately 1.3 million tons (equivalent to 45%) of generated waste are not collected.[1] The average collection rate in Senegal is 55%, which is lower than the average collection rate in sub-Saharan Africa (65%).

The rate of waste generation in Senegal is around 0.57 kg per day per person on average (national average). This rate is higher in larger and more industrialized urban centres such as Dakar, where the waste generation rate is about 0.8 kg per person per day. This rate is consistent with the average production rate in Sub-Saharan Africa

[1] World Bank (2017).

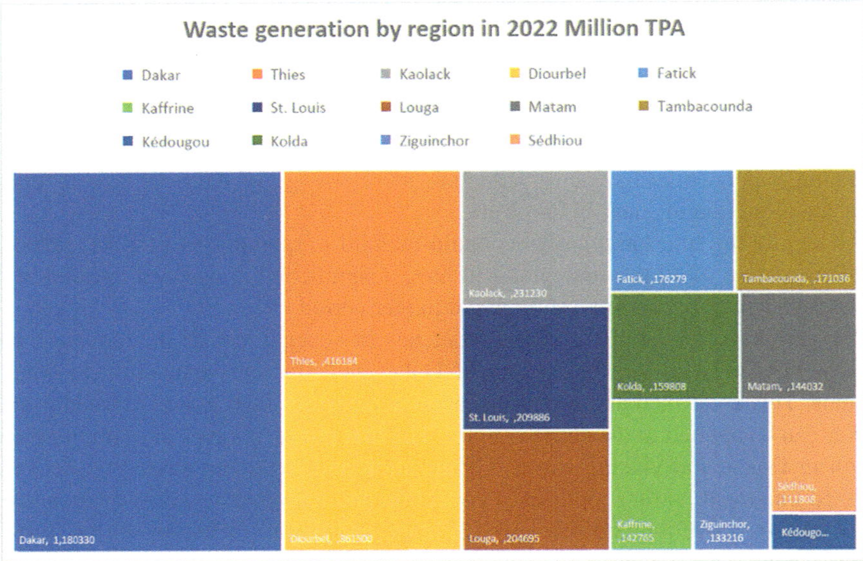

Fig. 4.2 Waste generation by the 14 regions of Senegal in 2022 (Million TPA)

(0.65 kg/person per day).[2] As shown in Fig. 4.2, the Dakar region has the highest annual waste generation of about 1.1 million ton followed by Thies and Diourbel. Most of the capital cities of the 14 regions have a waste potential generation of more than 5000 TPA.

Figure 4.3 provides pictorial representation comparing the waste generation in Senegal in 2022 and 2050. The waste generation in 2022 is calculated in accordance with the per capita waste generation of 500 g/day in Senegal on average and 800 g/day in Dakar, in accordance with the report SECTOR BRIEF SENEGAL: Solid Waste Management and Recycling, GIZ, 2021[3] and Senegal Municipal Solid Waste Management Project, World Bank, 2017.[4] Population for year 2050 has been projected by calculating 3% increases in population every year.[5]

The projected waste generation in 2050 is calculated in accordance with the population increase rate and per capita waste generation of 570 g/day in Senegal on average and 800 g/day in Dakar considering the national average per capita waste generation to increase from 500 to 570 g/day. With a population growth rate of approximately 3.0% per year in Senegal could generate 8.95 million TPA of waste by 2050. Considering the non-existence of a waste management system in Senegal, there is a greater need to intensify proper waste management practice and expand the capacity of waste management in Senegal.

---

[2] World Bank (2012).

[3] Sector Brief Senegal: Solid Waste Management and Recycling. GIZ, 2021.

[4] World Bank (2017).

[5] Population of Cities in Senegal (2023).

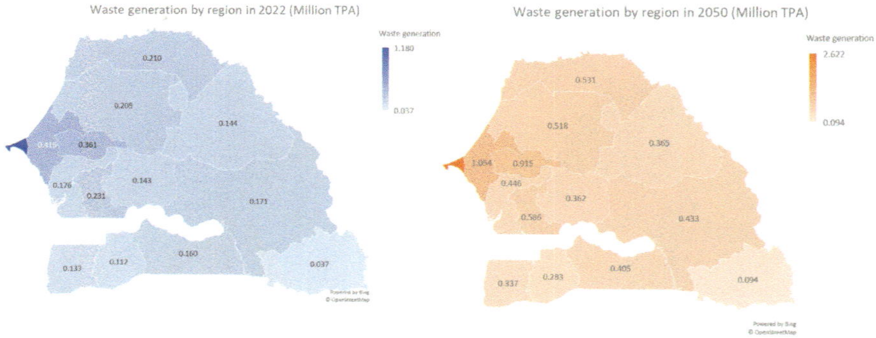

Fig. 4.3 Comparison of waste generation in 2022 and 2050

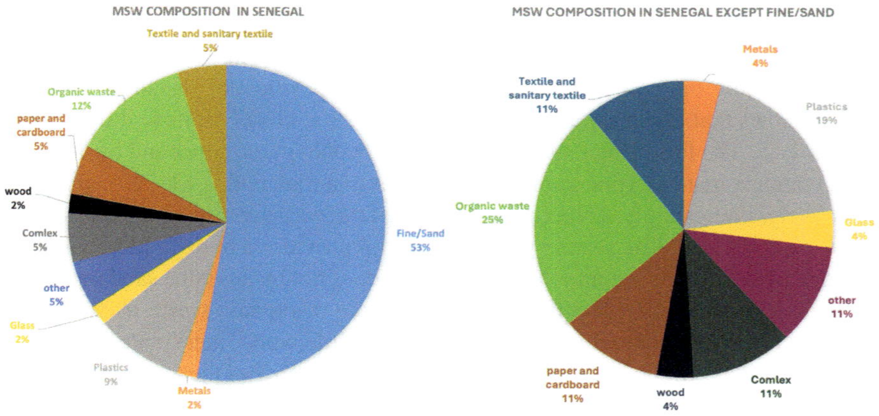

Fig. 4.4 Composition of waste in Senegal including and except fine sand material[6]

Sand, mud and organic debris make up the majority of Senegal's solid waste which contributes to the difficulty and expense of the country's waste management. The composition of solid waste in Senegal varies from large urbanized areas like Dakar to secondary cities, according to a study that was carried out nationwide by the Unite de Coordination et de Gestion (UCG) but in general it is made up of two thirds of fine particles, sand and organic waste (see Figs. 4.4 and 4.5). This is the case regardless of where the study was conducted. Only one third of the waste is composed of the other materials which include things like paper, cardboard, plastics and textiles among other things. The composition is comparable to that of other countries in Sub-Saharan Africa but it is strikingly dissimilar to the composition of waste in developed countries which drives up the cost of transportation and restricts the range of technologies that may be used to treat the waste.

---

[6] Sector Report Circular Economy Senegal (2022).

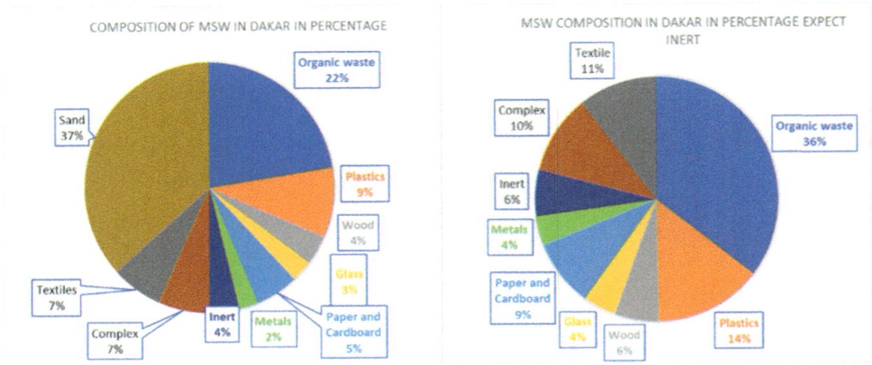

Fig. 4.5 Composition of waste in Dakar[7]

The composition of Senegal's waste stream is depicted in the informative pie charts that show the waste's composition. The composition of the first graph of Fig. 4.4 shows that organic waste predominates at 28%, closely followed by plastic waste at 21%. The following materials contribute: wood, glass, paper and cardboard, metal, complex material (fine sand material), inert material, and 5, 12, 12, 5, and 5%, respectively. Interestingly, a significant amount of the waste composition is made up of fine sand material. The second graph of Fig. 4.4 provides a more distinct view of the organic and inorganic components because it does not include fine sand material. The updated composition highlights the importance of organic waste in the waste stream by revealing a higher percentage of 33%. While wood, inert material, glass, paper and cardboard, and metal contribute 5, 14, 5, 14, and 5% of the waste generated, plastic waste continues to be a significant source at 24%. These graphs highlight how crucial it is to take fine sand material into account when having conversations about waste management because it has a big impact on Senegal's total waste composition.

The waste composition pie charts in Dakar offer important insights into the various material categories that contribute to the overall waste generated is shown in Fig. 4.5. The first graph of Fig. 4.5 shows a varied distribution when fine sand material is included. Notably, the largest portion is made up of fine sand material (37%), which is followed by organic waste (22%), plastics (9%), and other materials. The overall waste landscape is also influenced by the presence of wood, complex material, inert material, glass, paper and cardboard, metal, textiles, and sanitary materials. The second graph in Fig. 4.5, on the other hand, excludes fine sand material and concentrates on the composition of waste. The distribution percentages change noticeably as a result of this modification. With 36% of waste, organic waste leads the way, followed by plastic waste (14%), complex material (10%), and other waste. The removal of fine sand material emphasizes how important these categories have become in the waste stream. Furthermore, appropriate adjustments are made to the proportions of

[7] Solid Waste Management City Profile, Dakar, Senegal. dakar_city_profile_v2.pdf (ccacoalition.org). Retrieved on 05.05.2023.

wood, inert material, glass, paper and cardboard, metal, textiles, and sanitary materials. These graphs highlight how much fine sand material affects Dakar's total waste composition. A detailed understanding of the distinct contributions of different materials can be obtained by analysing the waste streams with and without fine sand material. This understanding serves as a basis for targeted waste management strategies and environmental interventions in the area.

Wastes from hospitals especially hazardous waste pose a significant risk to the natural environment. Even though there are licensed businesses that collect hazardous wastes, the final destination of the hazardous waste collected is unknown due to either no appropriate treatment facilities for these wastes or the facilities that are in place have become obsolete.[8] Incinerators designed specifically for the disposal of medical waste can be found in only few selected medical facilities. Because of this situation the hazardous waste is being disposed of without discrimination as part of municipal waste at the Mbeubeuss dumpsite and other dumpsites across the country. At these dumpsites thousands of people are working as informal recyclers without proper protective equipment which poses a serious risk to both human health and safety as well as the environment. The fact that hazardous waste and medical waste have been left without proper accounting is becoming an increasingly pressing issue for the sector.

Due to the variety of solid waste kinds each stream has considerable potential (Figs. 4.6 and 4.7). The potential rate of energy recovery from textiles, wood, and plastics is 24%. The potential rate of organic recovery from putrescible waste is 25%. The prospective rate of resource recovery by recycling (mostly iron and other metals, plastics, paper and cardboard and glass) is 29%. Note that all mentioned valuable materials represent approximately 50% of the total waste composition in Senegal, whereas the other 50% is sand which was taken out of the composition. Sand can be recycled in the beach zones better than using it in landfill sites (see footnote 3). Immediate waste management solutions are urgently needed in Senegal due to the increasing population growth rate of approximately 3.0%. By 2050 Senegal could generate up to 8.9 million tonnes of waste (4.45 million tonnes are sand and 4.45 million tonnes are valuable fractions). Currently, there is no waste management system in place and hazardous waste from hospitals is not properly managed causing severe damage to the environment and human health. Inadequate waste source separation practices, scattered waste collection, and lack of waste collection mediums are additional problems. There are no options for waste treatment at the household level and there is no method to treat organic waste and recyclable waste. Additionally, there aren't enough hygienic landfills in the nation to properly dispose of waste.

By increasing the percentage of waste to energy (waste incineration), a different scenario can be achieved. In this scenario, the incinerators can handle between 40 and 50% of the waste generated. As a result, the previously mentioned waste distribution will change, with the sanitary landfills receiving 15–20% of the waste instead of 63%, 12% going toward composting or biogas production, 7% going toward RDF production, and 18% going toward recyclables.

---

[8] World Bank (2017).

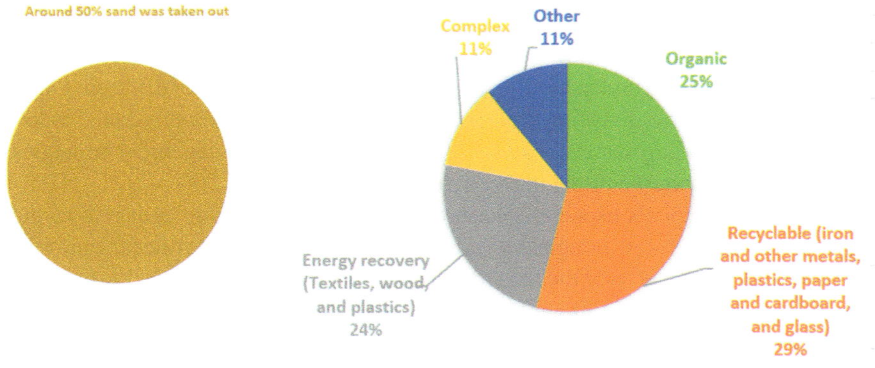

Fig. 4.6 Solid waste value potential of the generated waste in Senegal

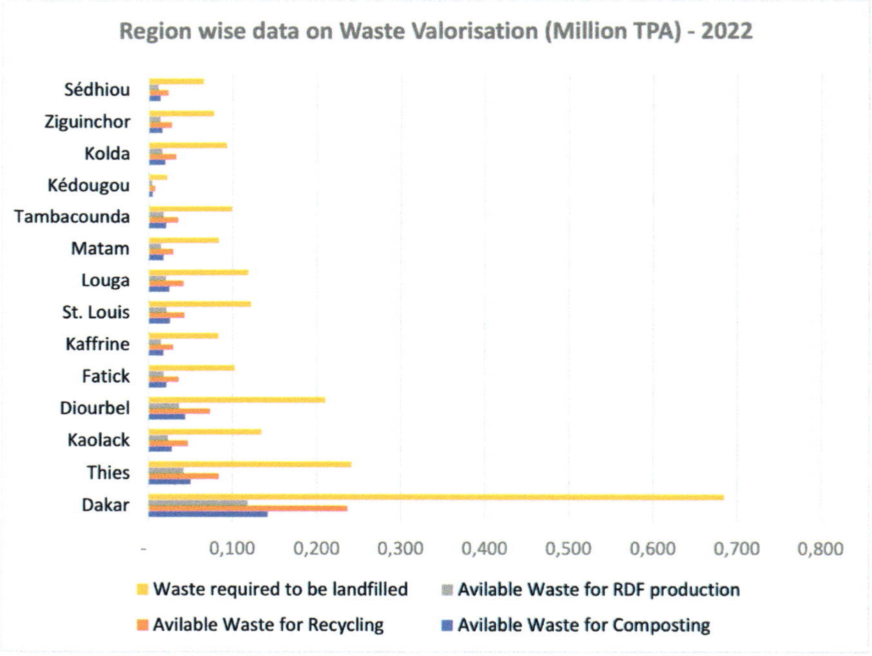

Fig. 4.7 Waste valorisation by regions in 2022 (Million TPA)

4.3 Energy Generation and Consumption in Senegal

Senegal is among the nations facing numerous obstacles and problems in the energy sector worldwide. Senegalese forests are under a great deal of pressure as a result of the country's heavy reliance on biomass as a primary energy source—more than 50% of the country's energy consumption comes from this source. It is important to note that using biomass for cooking contributes significantly to the depletion of the nation's forest stocks. Deforestation will, on average, result in shortages in the energy supply for cooking, especially for the poorest households.

Every year, more than 40,000 ha of forest are lost due to desertification, forest fires, and firewood harvesting. In addition, the cost of conventional fuels like charcoal and firewood has increased over the past ten years. Firewood provides Senegal with more than 45% of its basic energy needs (31% directly as firewood and 14% as charcoal). Stated differently, there exists a clear correlation among the energy sector, energy consumption, and deforestation. This implies that by incorporating alternative energy sources into the energy mix, not only will the energy requirements be satisfied, but the depletion of Senegalese forests and biomass will also be addressed, ultimately leading to a reduction and realization of CO_2 emissions.[9]

Senegal's energy industry is made up of 27 TWh of annual fossil fuel supply, or over 40% of the country's total energy contribution. The remaining energy is exclusively derived from biomass. The majority of domestic energy use is likewise accounted for by cooking. Only around 10% of the total energy used is used to produce electricity, which is mostly produced by fossil fuels (diesel generators) and a tiny amount of renewable energy (hydro, wind, and solar).[10]

In terms of electricity and power generation the annual increment rate in power generation is around 7% but the power demand grows a lot faster (314% since 2000) than its supply due to that and other technical reasons in the utility grid and lack of organizational structure make around 67% of the Senegalese people have access to the utility as Senegal is a country of 17.3 capita (2022) means that there are almost 6 million people without power access. In comparison to neighboring states, the electrification rate is relatively high: at 75% in urban areas and 17% rural, the national average is 45%, up from approximately 30% a decade ago. In comparison in neighbouring Mali national coverage is just 25%.[11]

A Poverty Reduction Strategy Papers were prepared by member countries in broad consultation with stakeholders and development partners, including the World Bank staff.[12] The main target of this strategy is to reduce poverty in Senegal main part of it is the development of the energy sector as a major driving force to overcome poverty. Ten objectives were stated four of which are essential for this work:

- diversify energy sources and technologies
- promote the control of energy and renewable energy sources
- provide better and safer public access to domestic fuels
- enhance access to energy services in rural and suburban areas so as to facilitate the working of basic infrastructures (schools, healthcare facilities, conservation infrastructures, etc.).

---

[9] "energypedia," 9 Septemeber 2020. [Online]. Available: https://energypedia.info/wiki/Senegal_Energy_Situation. [Accessed 27 11 2023].

[10] "afrik21," [Online]. Available: https://www.afrik21.africa/en/senegal-government-exempts-renewable-energy-equipment-from-vat/. [Accessed 27 11 2023].

[11] Senegal: Poverty Reduction Strategy Paper (2007).

[12] https://www.imf.org/external/np/prsp/prsp.aspx.

It can be seen from the objectives that the renewable energy sector should be extensively enhanced and accelerate its development. Further the diversity of energy sources and their related technologies emphasizes the importance of considering waste (mainly Municipal Solid Waste) as a source of energy (fuel) needed to be used in different manners and applying different technologies. Note that waste to energy concept plays two major roles; first is applying the concept of waste as a source of energy and second is matching the point of enhancing access to energy services in rural and suburban areas. It can be said that waste to energy concept can decentralize the energy generation for the rural areas by converting the daily generated waste by the inhabitants into useful energy.

As a main driving force to push forward in the renewable energy sector in Senegal and invest in this sector is the Senegalese Ministry of Oil and Energy has recently decided to exempt equipment for the production of renewable energies such as solar, wind and biogas from VAT (value added tax). The government is aimed at accelerating the electrification of countryside areas in Senegal. The government's decision is part of a much broader strategy aimed at achieving universal access to electricity in this West African country by 2025, where the Senegalese government is trusting decentralized solutions such as solar-mini grids, solar kits and large grid-connected power plants, which align with the concept of waste to energy that seeks to erect biogas plants, incinerators, and gasifiers in different capacities according to population density in each site for energy generation and hydrogen production as a mobile source of energy that can be transported everywhere inside Senegal.

References

Population of Cities in Senegal (2023) (worldpopulationreview.com) Retrieved: 16 May 2023
Sector Report Circular Economy Senegal (2022) Ref IUC: 202108079 / PST: PST21SN02
Senegal: Poverty Reduction Strategy Paper (2007) International Monetary Fund, Washington, D.C.
World Bank (2012) What a waste
World Bank (2017) Senegal municipal solid waste management project

Chapter 5
Strategy for Waste Minimization, Storage and Waste Management at Source

In an integrated waste management approach waste prevention and minimization are at the top of the hierarchy. Waste prevention implicates actions taken before a substance material or product becomes waste. Waste prevention actions result in the reductions of waste quantity generated through the re-use of products or the extension of the life span of products (quantitative prevention) and limit the adverse impacts of harmful substances on the environment and human health. Waste prevention could be accomplished by taking proactive steps like enhancing eco-design and manufacturing processes, as well as by encouraging the general public and private consumers to except durable, long-lasting products that are operated and maintained carefully. The advantage of waste prevention strategies is to overcome prevailing barriers such as lack of information, lack of resources and lack of awareness to improve efficient and profitable production and consumption. Waste prevention is most effective if undertaken during the design phase; reuse plays an important role in waste prevention but can lead to tenacious effects. Recycling and waste prevention are related but require different policy approaches.

5.1 Waste Minimization Through Animal Feeding

Food enterprises, including supermarkets, restaurants, cafeterias, and markets, play a significant role in simultaneously identifying solutions and cutting expenses. Figure 5.1 describes different waste management and prevention options for food businesses and markets.[1] Conducting an organic waste assessment is the first step in reducing the amount of organic waste generated. Such an assessment will help in understanding the exact organic waste stream that is being thrown away. Being aware of what exactly is being discarded there is a higher chance to reduce disposal costs,

[1] Eriksson (2015).

S. Narra et al., *Sustainable Waste Management in Senegal*, World Sustainability Series,
https://doi.org/10.1007/978-3-031-98968-1_5

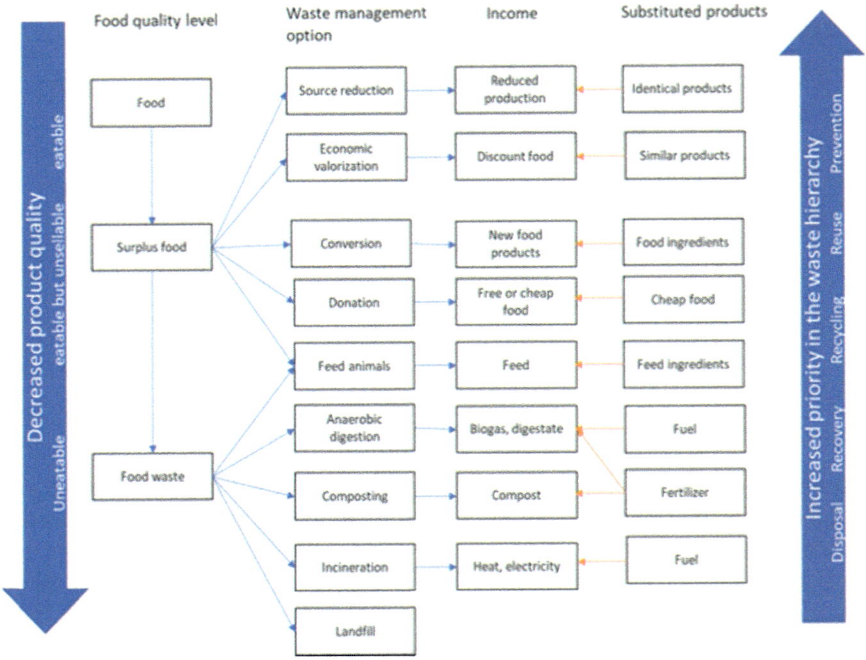

Fig. 5.1 Food waste management and prevention options for food businesses and markets

reduce over-purchasing and labour costs, reduce the amount of water and energy used for production and reduce greenhouse gas emissions.

After reducing the avoidable sources of organic waste a second focus has to be given to enhancing the reuse mechanism of generated organic waste (see Fig. 5.2). In Senegal animal feed systems are traditional methods of consuming organic waste including food waste. To promote the waste recycling scheme in regions of Senegal, the focus has to be given to enhancing the efficiency of the existing system. Food processing industries and centres can cooperate with piggeries or animal farms to ensure the effective collection of available food. Animal farms can also support the collection system financially. For example animal farms could contribute to the purchase of a collection fleet for separate food waste collection or provide financial support for the collection and delivery of organic waste.

An alternative option for towns in Senegal would be to support the existing system and officially link the animal farms with the commercial generators that could provide a mechanism for the collection of organic waste from different hotspots. Not only does the strategy and execution procedures depicted in Fig. 5.2 represent a long-term, sustainable solution for the management of organic waste produced by commercial generators, but they also decrease the burden on municipalities. In the longer phase the system can be improved when organic waste reusing schemes are well-defined.

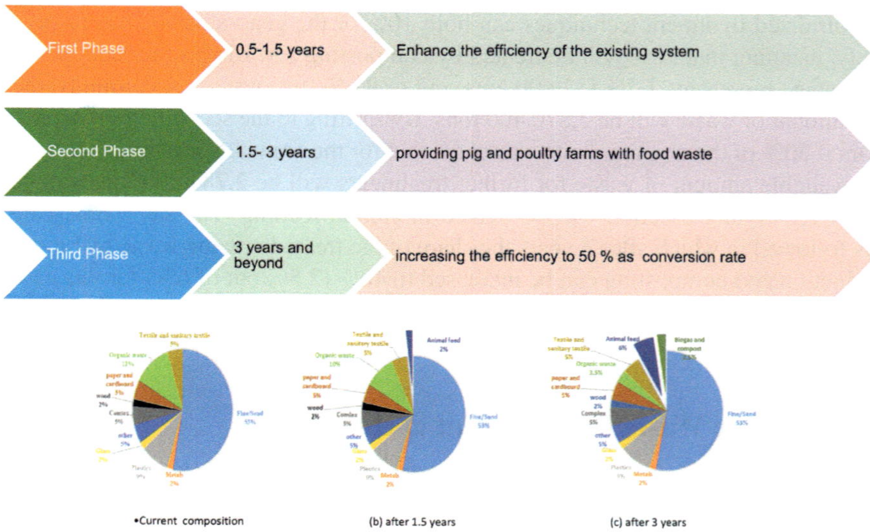

Fig. 5.2 Step-by-step enhancement of food waste management strategy (fraction of organic waste)

This could be achieved by considering practical and efficient collection and recycling systems and backing them up with the necessary bylaws to assure their implementation. A clear mechanism for interaction between municipalities and relevant stakeholders should be developed. In addition the municipalities should inspect environmental and hygiene-related matters of animal feeding in close cooperation with the Pollution Control Board committees.

For example, education on the safe reuse of food and compiling a manual that specifies which organic wastes are suitable for reuse. Equally crucial is the promotion of active public awareness campaigns through the dissemination of program objectives and opportunities for stakeholder engagement. It is essential to prioritize the enhancement of institutional arrangements, the annual assessment of goals and successes, and the evaluation of stakeholder responsibilities. In the section that follows, alternative strategies for promoting organic waste management are described.

The most important point that needs to be highlighted is that the proposed strategies have a direct effect on the composition of the waste, meaning that the waste composition will be changed step by step with time as a consequent effect of the applied strategy. Again, Senegal has generated around 3 million tonnes of waste in 2022. Organic fraction occupies 12% (360,000 tonnes organic). By assuming that a decentralised waste minimization strategy will be applied to treat 20% of the organic waste to be supplied for the animals' farms as food then the composition of the waste after 1.5 years (during the first phase) will be changed as follows: The composition of the organic waste will be changed to 9.6%,where 20% would be reused as a feeding stuff for the animal represents around 2.4% of 12%, therefore 9.6% will stay as an organic fraction as can be seen in Fig. 5.2. This means that 2,928,000 tonnes of waste will be available for further treatment. In the same phase, compost and Biogas as decentralized treatment techniques can hold 10% of the generated amount respectively, meaning that 20% will be treated by composting and Biogas production this percentage represents 1.2% (57,600 tons) out of 9.6% (rest waste), accordingly, the rest amount of waste will be 2,870,400 tons. By shifting to the second phase, after 3 years if 50% of the organic waste was converted for the animal feeding, in this case, the available amount of waste for further treatments will be 2,784,000 tons, where 6% of the organic fraction will be treated in animal feeding. The same timescale was followed in which after 3 years the share of the treated organic waste by biogas production and composting can be increased to 40% (2.5% out of 6%) for this case the available amount of waste will be 2,745,000 tons.

5.2 Organic Waste Management at the Source

Home composting is a simple method to treat organic waste at the household level. Practising the household composting method contributes to the diversion of organic waste from landfills/dumpsites and reduces the cost of SWM. The final product can return to the soil as much or as a soil conditioner. Home composting is an effective solution in small towns to the SWM dilemma. To promote the organic waste treatment at the household level in the first phase, a suitable home-composting method should be identified to be adapted locally. The benefits and drawbacks of each system are the major drivers in adapting a composting technique to the local context of the people of Senegal.

The capacity building comes in the second phase; the trainers should get training at the ward level on the requirement of each composting method and optimum operating conditions. These technical know-hows should be transported to the citizen for successful management of organic waste in homes. To achieve the latter a key factor is that in all types of home composters proper segregation of organic waste. Kitchen and garden waste can be used as raw input materials. However meats, bones, oil and dairy products should not be added to the system. The small volume containers are ideal to treat fruits, vegetables, food and garden leaves. To control the moisture in the container fruits like watermelon should be squeezed before addition and if possible the leachate should be collected. The smaller the particle size the faster the decomposition process. In containers where air circulation is restricted materials should be mixed regularly and bulky agents such as sawdust or dry leaves should be added. To accelerate the composting soil or cow dung can be added between layers. Another way to fix the available inoculum is to add the sawdust and additional materials to it. This would help to control moisture, reduce odour, increase porosity and speed up the process. Few recommendations on the most adopted composting models in Senegal as an example are further mentioned and demonstrated in Table 5.1.

Additional parameters also have to be considered for producing a good quality compost. For example, placing a composter container is a factor that affects the proper implementation of the process. Locating the composter container indoors or under a shelter is crucial to avoid direct sunlight and rainwater penetration. If the

Table 5.1 Composting techniques at a household level—benefits, drawbacks, and recommendations

TYPE	Benefits and drawbacks	Recommendation	Model
Plastic Bucket	Benefits: low-priced, suitable for small households, shorter decomposition time Drawbacks: needs daily monitoring, generates odour or attracts pest	Chopping the waste, regular mixing using a fork, addition of sawdust, soil, or spraying inoculum to speed up the process	 *Source* Atin et al. (2021)
Crate composting	Benefits: low-priced, suitable for small households, shorter decomposition time Drawbacks: temperature control, needs daily monitoring, generates odour or attracts pest	Using cardboard to line the inside of the crate can assist to maintain a consistent moisture level	
Pot	Benefits: low-priced, leachate collection outlet Drawbacks: low air circulation	Drilling holes in the pots for air circulation, the addition of saw dust, soil, or cow dung between layers	
Three bins	Benefits: light weight and easy to handle, inexpensive, able to start the process for fresh materials Drawbacks: no moisture and temperature control	Lining the bottom of buckets with newspaper, the addition of sawdust, soil, or cow dung between layers	
Khamba	Benefits: able to start the process for fresh materials, treating higher volume of organic waste Drawbacks: no moisture control	Lining the bottom of pots with newspaper, Drilling holes in the pots for air circulation, chopping the input material, addition of soil or cow dung to pots	

containers allow leachate collection, the excessive moisture can be collected over the period of the composting process. The collected leachate can be applied to the soil after diluting with a 1:10 ratio (leachate: water). The composted material can be removed from the container after 4–6 weeks according to the composting method. At the last stage as an indicator, the composting materials turn dark and smell like fresh earth. To purify the composted material the final product can be sieved and the coarse material can be returned to the container for composting again. The produced compost should be placed on the ground at least for two weeks to mature before application to the ground. The trainer must check household organic waste management on a regular basis in order to maintain system control and enhancement. Recommendation put out is that each town establish an operational design encompassing the various composting processes. The installation of such a functional model will provide the citizen with direct insight into the process. Establishing awareness and providing suitable capacity building is crucial for achieving success.

5.2.1 Decentralised Composting Facilities

Decentralized treatment of organic waste is inadequate in Senegal. The fact that this method eliminates the need to incur additional costs for waste collection and long-distance transport must be addressed. The reduced operational complexity and cheaper running expenses of the decentralized facility encourage community-based garbage management and generate employment for the residents. A community-scale composting facility eliminates the demand for additional disposal facilities, promotes organic waste recycling locally and creates small-scale businesses and enterprises.[2] In this regard first of all any existing systems should be optimized by troubleshooting the functional and non-functional decentralized composting facilities. To this end capacity building through training of trainers and integration of awareness-raising directly on storage bins and decentralized composting plants are crucial.

In order to obtain a high-quality final product effective design and management of the process is the key element. The frequency of waste collection should be considered according to the generation rate and weather conditions. Organic waste decomposition in the storage containers results in odour and leachate generation. It is also crucial to collect waste in separate fractions for avoiding impurities and contamination. The waste collector removes the unwanted material from the organic waste fraction. A manual sorting can be conducted on received feedstock in the decentralized facility. Figure 5.3 provides an overview of criteria and parameters (site selection, collection system, composting system, operation and marketing) that has to be considered in a community-based composting approach.

[2] Bruni et al. (2020).

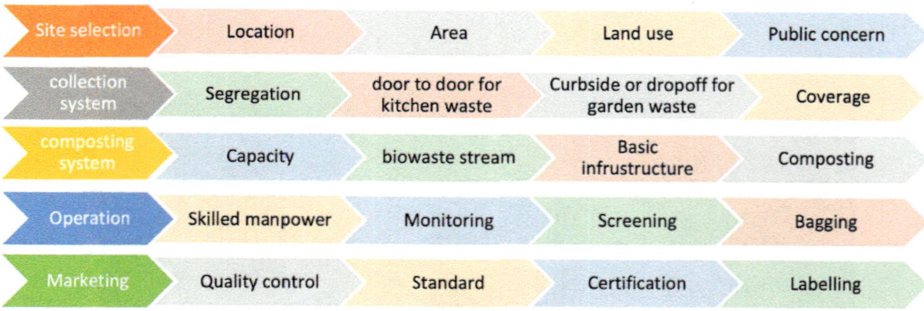

Fig. 5.3 Flow chart of the community-based composting approach

Site selection is a common condition that has a considerable impact on the development of a community-scale composting facility. The available site should be evaluated based on collection distance, accessible area, land use, community interest, existing organic waste stream and estimation of future organic waste generation trends. The capacity of the system has to be assessed based on the obtainable organic waste streams. Different methods for primary collection of bio-waste include door-to-door, block, curbside and communal collection. The door-to-door method is the most effective for source segregation but requires more resources. Block and curbside collection are used in communities with better cooperation. Once a collection system is ensured the choice of composting technology is the next step.[3] Several composting methods can be implemented on-site for communities, institutions and bulk generators of the waste. Economic resources, available materials, climatic conditions, technical infrastructures, vacant or available land space, waste volume and composition, physical and chemical properties of the waste, environmental considerations, and marketing requirements can all influence the selection of viable techniques (see Table 5.2).

To guarantee the success of decentralized composting facilities it is crucial to consider certain vital aspects. These include securing a reliable freshwater supply for the composting process, employing skilled workers, building a roof that regulates moisture during dry and rainy seasons, providing a sorting area with a concrete surface and a sorting table setting up a manual sieve for screening and establishing a storage area for packaging the finished products.[4] The key to a successful composting system is a strong market demand and the effective sales of high-quality compost. This can be achieved through rigorous quality control measures, obtaining proper certification and establishing a reputable brand.

The branding of the end-product is necessary to create trust for the farmers and communities. These efforts are crucial for ensuring successful sales and long-term growth in the industry. The rainwater can be collected during the wet season by designing a rooftop construction system. With a simple structured design using a PVC pipe at the edge of the top of the roof the rainwater can be collected in the storage

[3] Atin et al. (2021).

[4] Zurbrügg et al. (2005).

Table 5.2 Type of composting technique model, their benefits and drawback

Type	Benefits and drawbacks	Recommendation	Model
Pit composting (dug into the ground)	**Benefits**: low-priced the pit size is adjustable to feedstock **Drawbacks**: the complete process is non-scientific	2 m × 1 m area per each pit spreading cow dung slurry at the bottom, chopping the organic waste and covering with soil or brown material, in-situ monitoring	
High-density polyethylene (HDPE) container	**Benefits**: improved air circulation, fit for any area **Drawbacks**: temperature control, generating odour or attracting pest	Materials should be cut and mixed with dry leaves or sawdust. Should not contain any oily or liquid items. Water should be sprayed to maintain 50% moisture content	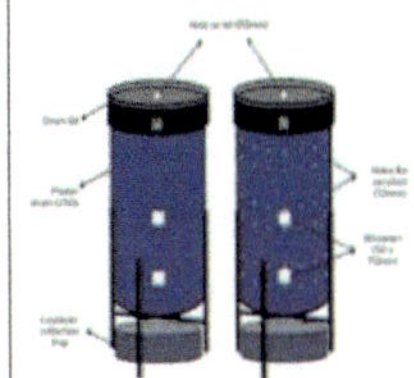 *Source* Atin et al. (2021)
Bin composting	**Benefits**: Inexpensive, fit for any area **Drawbacks**: air circulation, leachate control, generate odours	Materials should be cut and mixed with dry leaves or sawdust. The material should not contain any oily or liquid items. Frequent turning is required. The mechanism for collecting leachate from the bottom of the bin has to be adapted	

(continued)

Table 5.2 (continued)

Type	Benefits and drawbacks	Recommendation	Model
Pit composting: (constructed partially above and partially below ground)	**Benefits**: low-priced, potential to treat larger quantity of compost **Drawbacks**: require skilled manpower	Proper roofing, leachate collection, and wall aeration are crucial. Floor sloping aids leachate flow. Building a small pit next to composting units and connecting it with sloped pipes allows for more leachate collection	
Vermicomposting	**Benefits**: high-quality final product **Drawbacks**: require adequate area and skilled manpower, the higher final price	proper for the ward level, pre-composting and chopping the organic waste ease the process, cover the piles to keep the moisture	

tank which can be later used during the dry season. Composting material can attract animals and make a nuisance for the citizens; therefore it is highly recommended to fence the system set up. Ensuring the successful implementation of decentralized composting facilities is maintenance and monitoring of the system. To this end monitoring tours for community and decentralized composting plants using a traffic lights system to identify the hotspot of contamination should be planned. The community bins should allocate precisely to the property owner or bulk waste generator to give a warning in case of violating the rules and raising awareness. After several warnings and training penalties should be introduced.

5.2.2 Decentralised Biogas Facilities

Table 5.3 lists the benefits and drawbacks of the decentralized anaerobic digestion systems used to produce biogas. Biogas technologies are categorized as "wet" and "dry" procedures. In rural areas, anaerobic digestion (AD) has long been a popular technique since the biogas produced by this process is usually high in methane, requiring less clean-up. Biogas is created from organic waste through a process called anaerobic digestion, which occurs without oxygen. Methane and CO_2 are the primary components of the biogas that is produced (approximately 60%). The main uses of biogas are for cooking, heating, electricity generation, steam production, and fuel for vehicles. Notedly, 2.14 kW of power can be produced with 35% efficiency from 1 m^3 of biogas collected. By reducing the quantity of waste that needs to be disposed of in landfills, biogas facilities can also be a promising solution to the issues related to MSW. The four steps of the biogas generating process are hydrolysis, acidogenesis, acetogenesis, and methanogenesis. Depending on the type of feedstock, the process's rate-limiting stage is hydrolysis. Pre-treatment of the organic wastes is necessary to reduce the digestion time and increase the output of product gas.

5.3 Approach for Promoting Waste Minimization and Waste Management at Source

Following is the approach for promoting alternative organic waste management techniques in Senegal that have to be divided into short-, medium- and long-term goals.

Short term goals (up to 1 year): The importance has to be given to finding an approach to reduce the amount of organic waste being generated, reuse mechanisms for generated organic waste whenever possible and develop standard treatment techniques for different organic waste produced in Senegal. Within the short term goal the primary focus has to be given to reducing the amount of organic waste generated. Food businesses like grocery stores, restaurants, cafeterias and markets have a major

Table 5.3 Type of decentralised biogas technique model, their benefits, and drawback

TYPE	Benefits and drawbacks	Recommendation	Model
Fixed-dome digester	**Advantages** – Highly cost-effective—relatively low construction cost – Comparatively expanded lifespan when constructed by a skilled worker – Due to the absence of moving and metal parts, less maintenance is needed – Less temperature fluctuation and saves space—as the dome is constructed underground – Capacity building and employment for skilled local workers **Disadvantages** – Skilled technical work is required to ensure the air-tight construction of the dome – The major concern of gas leaks when constructed by inexperienced masons – Inconsistent gas pressure depends on the volume of collected gas – In the event of problems occurring after construction, it is highly difficult to repair as the structure is located underground	Fixed-dome biogas plant comprises a closed immovable dome-shaped digester with a gas holder, a fixed feedstock inlet chamber and a displacement pit as an outlet to collect the digestate. These digesters are shaped like a dome and are usually built underground as seen in figure[a] Fixed dome digesters may be constructed using either concrete bricks mortar or a combination of both. They are the most commonly used digesters for household biogas plants in many countries in Asia and Africa	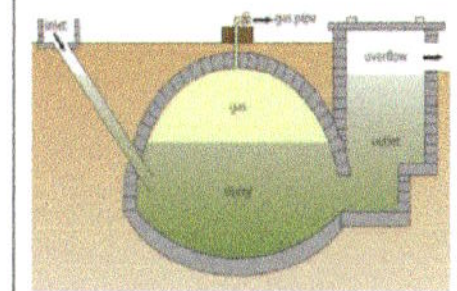 *A standard diagram of a fixed-dome biogas reactor*  Underground dome *(Deenbandhu) plant being built in South India (courtesy: Kingdombio)*

(continued)

Table 5.3 (continued)

TYPE	Benefits and drawbacks	Recommendation	Model
Lightweight Portable Bio Gas Plant	**Advantages** – Significantly less time is required for the complete installation of the plant – Easy to relocate whenever required – No special skills are required for the installation of the plant – Excavation of the earth is not required – Suitable for Households with a confined land area (as there is also a possibility to install on the rooftop) – These plants have better resistance to salty water and are suitable for places close to sea-level[b] **Disadvantage** – Susceptible to weather conditions during colder seasons—as the plant is installed above ground – Comparatively less biogas yield as plastic retains less heat compared to conventional plants constructed with concrete or metal[c]	This lightweight Portable Bio Gas Plant consists of a fermenter tank and on the top is a reserve tank (floating dome) called a gas holder to collect the produced biogas as shown in figure[d]. This lightweight Portable Biogas plant varies from a size of 0.5–6 m^3. Usually, the smaller units (0.5–2 m^3) are equipped with or without a water jacket, however, the bigger ones (> 2 m^3) are designed with a water jacket. In this technology, the gasholder does not float directly on the effluent but in a filled water compartment which is designed to minimize gas loss and improve the cleanliness of the biogas unit[e]	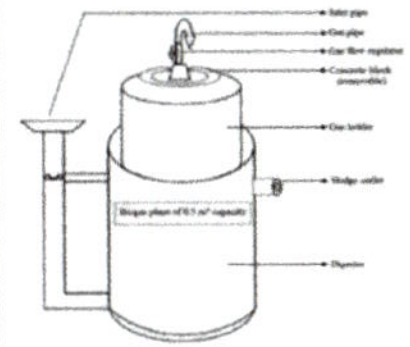 *Schematic sketch of a portable floating drum biogas plant* (Mohammed Shejir et al., 2017)  *Floating drum biogas plants in India*

(continued)

Table 5.3 (continued)

TYPE	Benefits and drawbacks	Recommendation	Model
Floating drum biogas	**Advantages** – Operation and design are simple – The floating drum directly visualizes the volume of the stored biogas – Possibility to regulate the gas pressure output using the additional weight – Relatively easy for construction and there is less possibility of problems concerning the construction error **Disadvantage** – The utilization of steel drums makes the construction cost high – Shorter lifespan – A floating drum is susceptible to corrosion when not maintained properly – Relatively high maintenance cost (de-rusting and regular painting of metal-drum is essential to ensure longer lifespan)	The floating drum biogas plant consists of a cylindrical digester and a movable floating gas-holder, with a digester size typically ranging up to 50 m^3. The digester section of the reactor is usually constructed underground with concrete, bricks, or quarry stone and then plastered as seen in figure	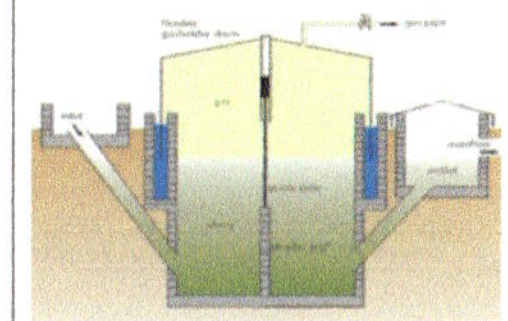 *Schematic diagram of a floating drum biogas digester*  *A picture from India of an FDB plant*

(continued)

Table 5.3 (continued)

TYPE	Benefits and drawbacks	Recommendation	Model
Continuous Stirred Tank Reactors (CSTR)	**Advantages** – Greater flexibility over the feedstock choice and better control over the complete process – Also suitable for high protein content waste – Low operation cost and higher efficiency – Optimum heat and feedstock distribution inside the digester **Disadvantage** – High investment costs – Frequent maintenance is required – Technically skilled personnel are required for the smooth operation of the plant – Comparatively less biogas yield as plastic retains less heat compared to conventional plants constructed with concrete or metal[f]	A complete Mixed Digester, or Continuous Stirred Tank Reactor (CSTR), is a biogas plant with a vertical steel or concrete cylinder tank covered with an impermeable cover to collect biogas. It is widely used for large-scale anaerobic digestion[g] Typically CSTRs are built above ground with the reactor volume > 130 m^3 digester and could be operated as a mesophilic or thermophilic reactor. Depending on the reaction temperature the retention time of AD can be as low as 15 days	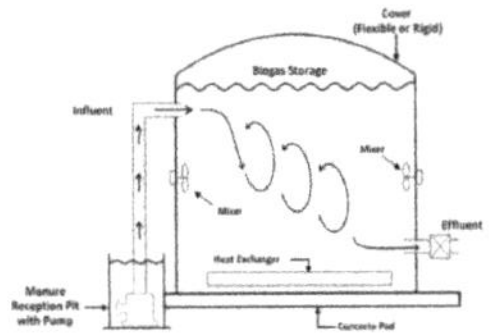 *Diagram of a completely mixed digester (Global Methane Initiative)*

[a] www.kingdombio.com/develop.htm
[b] Evaluation of small-scale biogas systems for the treatment of faeces and kitchen waste—Eawag, aquatic research
[c] Menkiti Nnamdi I, Ndirika Victor—comparative evaluation of fiber-glass reinforced plastic and metal biogas digesters
[d] Mohammed Shejir et al. (2017)
[e] Decentralised anaerobic digestion of market waste—case study in Thiruvananthapuram, India—Eawag, aquatic research
[f] Menkiti Nnamdi I, Ndirika Victor—comparative evaluation of fiber-glass reinforced plastic and metal biogas digesters
[g] Market opportunities for anaerobic digestion of livestock and agro-industrial waste in India, the global methane initiative Section 4 technology options (2020)

role to play in finding appropriate solutions and reducing costs at the same time. Conducting an organic waste assessment is the first step in reducing the amount of organic waste generated is necessary. Such assessment will help in understanding the exact organic waste stream that is being thrown away. Being aware of what exactly is being discarded, there is a higher chance to reduce disposal costs, reduce over-purchasing and labour costs, reduce the amount of water and energy used for production and reducing greenhouse gas emissions. After reducing the avoidable sources of organic waste, a second focus has to be created enhancing the reuse mechanism of generated organic waste. To promote the waste recycling scheme across Senegal the focus has to be given to enhancing the efficiency of the existing system. As a final step in the short-term goal a standard treatment technique has to be developed for the organic waste which cannot be avoided or reused directly. In this concern special waste streams such as coconut waste, water hyacinth, slaughterhouse waste, etc. with higher upscaling potential have to be addressed in particular and a tailor-made treatment technique has to be adopted for the individual waste streams. By-laws have to be updated favouring the Bulk Waste Generators (BWGs) who are attempting to manage organic waste at the source. Examples of programs that could encourage the management of organic waste at source include lowering the user charge for BWGs managing their own organic waste at source or waiving the user fee for BWGs managing organic waste at source for a period of two months each year. Finally, the waste streams without any upcycling potentials have to be utilised for nutrient recovery (composting) and energy recovery (biogas) and there could be a subsidy scheme for promoting successful organic waste management techniques such as composting or biogas.

Medium-term goals (2nd–3rd year): The medium-term goals start from the second year of action plan implementation and continue until the beginning of the third year. During the medium-term goal, importance has to be given to enforcing the standards of treatment techniques and market creation of the products. The tailor-made treatment technique concerning special waste streams has to be legislated. Such legislation ensures the quality and marketability of the product produced. Least but not last introductory of training for interested parties has to be conducted to provide information on technical, operational and maintenance details on the treatment technology (Biogas/composting). In addition importance also has to be given to appointing and training sanitary inspectors as a part of the medium-term goal. Appointing a sanitary inspector could be a crucial step in the medium-term goal. The sanitary inspector should have a proper scientific knowledge of entire organic waste management (if necessary training has to be provided to sanitary inspector specifically on composting/ biogas process) and should be capable of training the employees and public with the technical, operation and maintenance aspects of composting and biogas units whenever and wherever required. Senegal communities should also have functional biogas and composting units that are specifically designated for demonstration and education purposes and organize periodical training and enhancements to the trainers. The medium-term goals should also ensure the market for the produced product is fully developed.

Long term goals (3rd year onwards): Long term goals should mainly focus on the sustainability of newly adapted policy and by-laws concerning organic waste management at source. Importance has to be given to the replication of the standard treatment facility. In this respect funding schemes and handholding activities have to be provided for private parties to enable the sustainable growth of the sector. In this concern monitoring the installed composting/biogas unit and providing frequent training on operation and maintenance aspects would be a key factor that decides the success rate of the adopted organic waste management technology. The long-term goals should also focus on the collaboration between BWGs to ensure continuous availability of feedstock for smooth functioning of installed composting/biogas units. For the commercial units producing less amount of organic waste there could be a scheme for collaboration between the commercial units to ensure proper supply of the organic waste and smooth running of the installed treatment unit. The mid-term and long-term goals and achievements have to be monitored regularly to ensure a greater success rate.

This strategy and it implementation steps (see Fig. 5.4) has its specific designated benefits to enhance the MSW especially organic waste system of Senegal. The implementation of this strategy will also have an interlinking effect between the strategies that enable the management of organic waste in Senegal in the broader context. For instance as mentioned above direct effects of promoting of biogas technology in communities can not only aid in the management of organic waste but significantly influence the social-economical and environmental aspects especially in the rural areas. In a broader aspect the successful implementation of the above strategy will have additional benefits viz. (1) decrease the organic waste dumped at waste vulnerable points (2) decrease the load on the composting system (3) increase revenue generation (4) creation of more jobs (5) reduction in GHG emissions etc. The interlinking effects could be hard to quantify and the successful implementation of all the strategies will have a significantly increased interlinking benefit that could be seen in a shorter and longer period.

5.3.1 Improving Waste Storage at the Household Level

The expected categories of waste at the household level are organic, dry and hazardous domestic waste. This waste can be easily stored in separated bins with their allocated labels of coloured bins. UCG/SONAGEDs have been distributing storage bins to the citizens in residential areas. This approach is not sustainable for organic waste in the long terms and more emphasis should be given on organic waste management at source. In short term to ensure effective storage of waste at household level the schedule for the distribution of dustbins in different areas should run orderly and include the rural areas. Further the training and monitoring should be performed periodically to ensure the right application for the bins. To reduce plastic waste in general it is not advisable to use plastic bags in the organic waste bins. The organic waste bins can be brought to a water rinse daily. This also promotes the reduction of

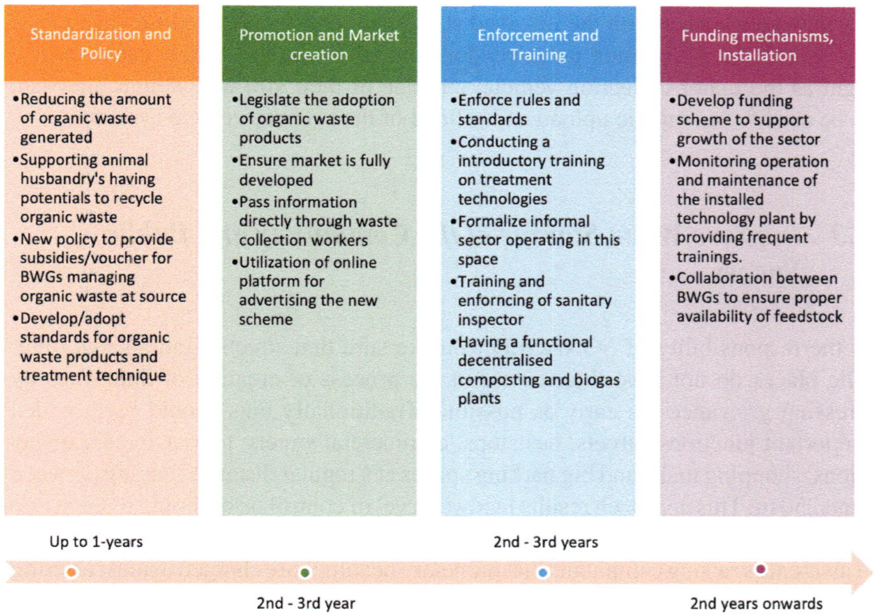

Fig. 5.4 Approach for promoting waste minimization and waste management at Source

plastic as a contaminant for any further processing. On the hand dry or recyclable waste collection in the densely populated households i.e. apartment complexes and cooperative societies can also be done in a simplified manner. For this a provision of a dry-waste-collection unit at a common place can be set up.

5.3.2 Improve Waste Storage at the Commercial Level

SONAGED can define the generation of a waste unit to declare a point source as a 'bulk waste generator'. A common definition is a generation of waste of more than 100kgs. Commercial spaces like supermarkets shops and businesses must segregate and store their waste in their designated places. Vendors and dealers of fruits, vegetables, and flowers should gather their waste in one location for additional processing or collection, and they should have a storage container designated for recyclable waste. SONAGED is responsible for the coordination and communication of these tasks between different sellers. Because the amount of bulk garbage generated varies and is uncontrollable based on the season and activities, it is recommended to set bins with a capacity of 200% of the projected bulk waste produced. A proper plan has to be developed together with the involved stakeholders. Bins installed for cumulative waste collection at common point vendors should be the same size, capacity, and weight as the major collection vehicle, and due of their size and weight, they can also be used for automatic upload and offload of the primary vehicle system.

5.3.3 Improve Waste Storage at the Community and Public Spaces

It is the responsibility of SONAGED to make sure that streets, stations and other public places do not have litter or are in the process of creation of dumpsites by addressing grievances as early as possible. Traditionally bins should be provided at important junctions, streets, bus stops, commercial streets, tourist spots, railway stations, shopping malls and big parking spaces at a regular distance ranging between 25 and 250 m. This approach results in lower level of control on dumping of the waste especially in the municipalities. The citizens are informed through collection works and awareness raising campaigns that monetary penalties are charged against littering or dumping the waste in public places. Where storage containers are installed all the bins should be marked, labelled or coloured at a consistency level across the Senegal helpful for raising awareness. Simple storages bins are more compatible with Senegal condition rather than modern systems. Figure 5.5 simplifies the waste storage actions needed per each settlement level of waste generation and collection to be considered in Senegal.

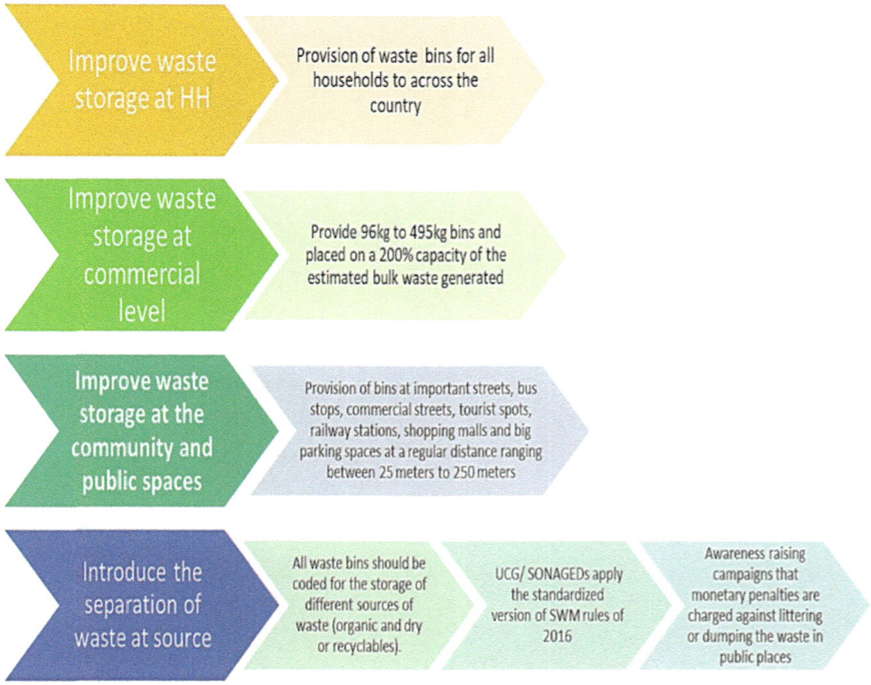

Fig. 5.5 Implementing strategy for closing the collection the gaps in collection of MSW

References

Atin B, Shailshree T, Subhasish P (Hg.) (2021) Decentralized-management-of-segregated-organic-waste. Centre for Science and Environment, New Delhi

Bruni C, Akyol Ç, Cipolletta G, Eusebi AL, Caniani D, Masi S et al (2020) Decentralized community composting: past, present and future aspects of Italy

Eriksson M (2015) Doctral thesis: supermarket food waste—prevention and management with the focus on reduced waste for reduced carbon footprint. https://doi.org/10.13140/RG.2.1.2502.3447

Mohammed Shejir R, George S, Anil KS, Abeena B, Sherin KG, Sunanda C (2017) Comparative study of the quantity and composition of biogas production using cattle, buffalo and goat excreta. Int J Livestock Res 36(12):2277–1964. https://doi.org/10.5455/ijlr.20170718035457

Zurbrügg C, Drescher S, Rytz I, Sinha AHMdM, Enayetullah I (2005) Decentralised composting in Bangladesh, a win-win situation for all stakeholders

Chapter 6
Strategy for Closing the Gap in Collected MSW and Transportation

6.1 Current State of MSW Collection and Transportation

The techniques and coverage rates of Senegal's solid waste collection and transportation services vary by region, and the organization in charge of collecting waste. This could be influenced by a number of factors, including the size of the area, the amount of manpower and money available, and most importantly, the condition and existence of a road network. These collection and transportation services are only available and effective in the regional capitals and in some departmental capitals. Moreover other small municipalities are struggling with the support from the Solid Waste Management Coordinating Unit (UCG/SONAGED).[1] The objectives of previously UCG/currently SONAGED were to support local and regional authorities in their waste management responsibilities by creating human and operational capacities within these communities. SONAGED's involvement in any municipality is case-dependent meaning SONAGED supports municipalities that lack collection transport and human resources with their own means. The Dakar regions waste management is however entirely the responsibility of the SONAGED.

The collection of waste services are provided by both informal and formal sectors providing primary collection to local neighborhoods or on a larger scale providing either secondary collection or an integrated collection service across the capital city. In Dakar waste management is integrative. The state local authorities (CADAK, CAR) and private authorities (Veolia) are part of the upstream while the local recyclers make up the downstream.[2] Each household in a high-income resident in cities has a waste bin and the collection vehicles like crushers or tractors equipped with boxes travel at a scheduled route stopping periodically to empty the waste bins of residents. The staff in charge of the collection is responsible only for the lifting and emptying of the bins placed in front of the houses. This system of collections occurs

---

[1] Kaza et al. (2018).

[2] Netherlands Enterprise Agency (RVO) (2022).

S. Narra et al., *Sustainable Waste Management in Senegal*, World Sustainability Series,
https://doi.org/10.1007/978-3-031-98968-1_6

only within areas with low population density and access to good roads at a fee paid by the residents based on a weekly frequency.

In high-population density and low-income areas informal collectors travel with wagons tricycles or horse carts through neighbourhoods and offer their services for a daily or monthly fee. Additionally central communal containers for the collection of waste are positioned at a street corner where users of around 10 or more households are located. Market areas are also part of this arrangement where people around the catchment area deposit their waste in communal containers. The frequency of collection of the communal containers is normally lifted at a daily routine or defined frequency. However waste generators in this situation do not pay any user fee; at least one waste staff is assigned to a central communal waste collection point to ensure the emptying of the waste-cans into the central containers. Likewise roaming waste pickers go door-to-door to collect recoverable materials from households and they usually focused on one material, e.g. iron, plastic or aluminium. The collection of recoverable materials can be structured into three categories; firstly the primary collection by many small waste pickers who can be surface technicians itinerant waste pickers or working on landfills; secondary collection by early acquirers who are often local wholesalers; and lastly tertiary collection distinguishing between industrial units and major exporters of raw materials. The collected plastic wastes are sold to manufacturers at fixed prices at collection points set up for individuals to deposit and sell plastics.

There is no efficient collection nor selective treatment of household or industrial waste across regions in Senegal. Senegal's capital, Dakar, has the highest collection coverage rate at 84 percent; another major city, Rufisque, comes in second at about 36 percent. At 33% in regional capitals, the collection rate is far lower than in secondary towns, which only reach 22%.[3] In Dakar city all waste collected from residents and surroundings is sent to transfer stations and collection vehicles on an average of two trips a day carry waste from transfer stations and communal collection points to the Mbeubuess disposal site.

However regional challenges such as scarcely urbanized areas and roads too sandy or narrow for vehicles cause inaccessibility of some areas resulting in inadequate and scattered service in Dakar and other capital cities. The existing collection services are irregular and unreliable. With no sanitary landfill, open dumping and burning are predominant in Senegal's secondary municipalities. The Waste Coordination Unit in Dakar also introduced a web-based monitoring system to communicate with citizens and optimized waste collection routes.[4] In Dakar the capital city major healthcare facilities have a secure central waste storage; moreover there is almost no indication of external transportation of Biomedical waste management (BMW) in Dakar. Yet it is recommended that the transportation of BMW outside of the health-care facilities must be done ideally with specific vehicles for this task and designed with an easy unloading, cleaning and disinfection system.[5]

---

[3] UCG (2016)

[4] Silpa Kaza (2018).

[5] Cheikh Dieng (2020).

6.1.1 Drawbacks

Inadequate waste storage bins: Annually more than 400,000 households in urban areas need regulation bins and the local authorities require more than 10,000 street litterbins and waste containers.[6]

Inadequate ineffective and scattered waste collection: Services in Dakar and other regional capital cities is due to scarcely urbanized areas and roads too sandy or narrow for vehicles causing inaccessibility in neighbourhoods. Additionally lack of waste collection mediums; more than 700 primary waste handling vehicles 750 waste containers and 850 waste collection vehicles are needed at a municipal level throughout Senegal's regional capitals.[7] Many municipalities such as Ngor that has no longer have space to position communal containers or assembly points hence the absence of waste ground results in illegal dumping.

There is no proper collection and transportation option for sanitary or hazardous waste: there is no separation between hazardous and non-hazardous wastes generated from medical and industrial sectors. The significantly high amount of hazardous waste generated is currently being either dumped or buried or burnt uncontrollably. The mismanaged hazardous and sanitary wastes can enter the organic waste stream; proper handling of these wastes is crucial to promote of hygiene and disease control.

No waste is weighed during collection: Waste from commercial generators is collected depending on the size and quantity of the waste. Yet there is no proper mechanism to weigh the waste that is collected from the waste collector.

6.2 Adapt the Proper Waste Collection System

There are several adapted methods for the primary collection of bio-waste namely: door-to-door, block, Curbside and communal collection. The door-to-door method is the most effective method for source segregation of organic waste. The door-to-door collection is especially useful in the areas having narrow streets where a collection truck cannot reach the individual house.

In the door-to-door collection system households can either place the filled containers outside their doors or hand them over to the waste collector upon their arrival. The waste received from the households will be directly emptied into the waste collection vehicle. The door-to-door waste collected can either be stored in a secondary collection point or directly transferred to secondary collection vehicles. The choice between secondary storage or direct transfer to secondary collection vehicles will depend on the availability of secondary collection vehicles, the area

[6] https://www.giz.de/en/downloads/SectorBrief_Senegal_Waste.pdf.

[7] Cabral et al. (2012).

of collection and the timing of collection. The requirement for secondary collection vehicles would be much higher if waste from all residential areas were collected during the mornings and transferred directly from primary to secondary collection vehicles in comparison to staggered timings of waste collection. The solution here would be to have a staggered timing of waste collection or have a secondary collection point where door-to-door waste collected can be stored before transferring it to a secondary vehicle. The wet dry and domestic hazardous waste must be transported in a segregated manner.

The door-to-door collection option is also the most effective method to raise awareness and promote segregation at the source. Despite having great advantages, door-to-door collection requires a higher labour force, time and investment as well as the presence of a household member during collection time.

In the areas with wider streets where the collection trucks can pass through conveniently a block or Curbside collection scheme should be implemented. Block collection can be implemented in the less populated areas where households can be easily informed by ringing a bell that the collection vehicle is arrived and is willing to collect the waste. The collection vehicles arrive at a specific location at a given time and date to collect household waste. Households bring their waste containers and empty them directly into the vehicle. In the case of higher population areas with wider streets the curbside method is more effective. The house owners leave the waste containers at the edge of the road. A waste collector removes the waste from the curbside or empties the containers into the vehicle as it passes through the street at a scheduled time and the containers are returned to their original place. In the curbside method fixing the schedule plays a key role in the proper collection of organic waste. Waste collection during the daytime e.g. 10 am to 2 pm is proved to be the most appropriate time for handling organic waste.

Particularly in the areas with significantly narrow roads utilization of communal bins could be highly cost-effective. In the communal bin method of waste collection depending on the waste generation rate a container is placed in a particular ward. In this method the waste generators bring the waste to the designated container and discard their waste. The content of the container is emptied daily into the collection vehicle and then the empty container is returned to its original place. The communal bin system provides greater advantages by reducing the number of labour required for the collection of waste from the areas with narrow roads that larger waste collection vehicles cannot reach. The disadvantage of the communal bin system is that if the waste is pressed into the huge bins and stored for a long time may cause leachate and odour nuisances. This problem could be overcome by appointing a person to monitor the community bins constantly. There is a visual comparison between the door-to-door and community bin collection system is shown in Fig. 6.1 which Senegal will adapt both systems depending on the type of street in each communities.

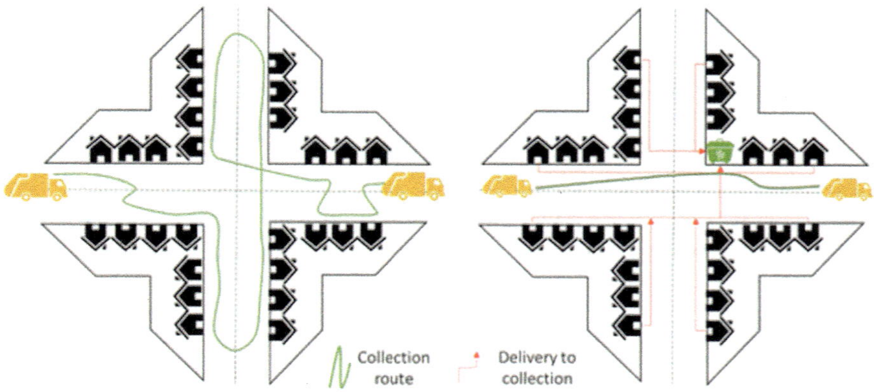

Fig. 6.1 Comparison between the door-to-door and community bin collection system

6.2.1 Select the Appropriate Type and Number of the Waste Collection Vehicle

The selection of a proper waste vehicle type and waste collector teams is another important aspect of collection service. The type of collection vehicle is dependent on various aspects which include population density, road size, financial liability and waste volume. During the selection of the transportation fleet prime importance has to be given to the separate storage containers in these vehicles. These containers should be used for separately transporting organic, dry and hazardous material. This would increase the value of waste. The waste workers should get trained on these requirements and they can be integrated into the collection vehicles. The same concept should be applied to the other waste stream; separate collection containers should be used for distinguishing the sanitary and hazardous waste from the rest of the recyclable waste. Handcarts/tricycle vehicles require low investment and they are appropriate for door-to-door organic waste collection on narrow streets of different areas of Senegal. Auto Rickshaws and tipper trucks require medium to high investment costs whereas compactor trucks need the highest cost. The tipper trucks and compactor trucks need wider space for collection service. All collection vehicles must consist of separate storage containers and must be purchased from reputed manufacturers to increase the lifespan and lower maintenance costs. The overview of the waste collection fleet is given in Table 6.1.

Once the type of collection vehicles is chosen, the number of collection fleets is estimated considering the density and number of households, waste generation rate, collection model (door to door, block, Curbside), frequency, number of staff, the required time for each trip and distance to the treatment facility. These factors are crucial for offering 100% coverage of organic waste collection services. Additionally integrating digitalization into the waste collection system can significantly optimize

Table 6.1 Overview of waste collection fleet

Parameter	Handcart 	Tricycles 	Auto Rickshaw 	Tipper Truck 	Compactor Truck 
Range	< 2 km	< 5 km	< 10 km	Unlimited	Unlimited
Road size suitability	Narrow	Narrow	Narrow	Medium	Wide
Volume	0.5 m^3	2 m^3	3–4 m^3	10 m^3	12 m^3
Labor requirements	1 collector	1 driver 1 collector	1 driver 1 collector	1 driver 2 collectors	1 driver 2 collectors
Cost	Very low	Low	Medium	High	Very high
Service life	5 years	5 years	10 years	5 years	5 years
Trips/day	2	2	3	4	3

the complete collection mechanism (Geo-tagging households and other waste generators, digitalization of money collection from households and commercial generators and their Tracking, monitoring, planning and management of vehicles/resource).

6.2.2 *Estimation of Collection Vehicle and Labour Required for Waste Collection*

Handcarts, tricycles and auto rickshaws can be used for the collection of waste in the areas having narrow streets. Handcarts and tricycles are particularly adaptable in the areas with alley ways of width < 2 m. Considering the capacity of a handcart as 0.5 m^3, the average waste generation per household at 2 kg, and the waste density at 350 kg/m^3 in a single trip waste collector having a handcart can visit ~ 90 households. With a maximum of two trips per day, a single waste collector with a handcart can visit about 180–200 households. With this calculation about 5–6 handcarts with accompanying labour are required for the collection of waste from 1000 households.

Considering the utilization of a tricycle with a 1 m^3 loading volume; a single trip could cover ~ 180 households. With a maximum of two trips per day, a tricycle with one rider/collector can visit about ~ 360 households. With the utilization of handcarts and tricycles there is a greater need for a secondary collection point where the waste collected from households can be temporarily stored till the compactor trucks collect and transfer it to the respective treatment facility. A single trip of a compactor truck of 12 m^3 loading volume is enough to collect the waste generated by ~ 2000 households.

Auto rickshaws can be used in the areas with alley-ways of width between 2 and 3 m. considering the utilization of an auto-rickshaw with a 3–4 m^3 loading volume;

a single trip with one driver and one collector could cover ~ 700 households. With a maximum of three trips per day an auto-rickshaw with one driver and one collector can visit about ~ 2000 households. When the transfer distance to treatment centres is short, auto-rickshaws can directly transport the collected waste to treatment centres without the requirement of additional secondary collection points and vehicles.

Tripper trucks are greatly suitable for a block or curbside collection in the areas with wider streets. Auto-rickshaws can also be used in the block and curbside collection with less populated areas. Considering the utilization of a tipper truck with a $10m^3$ loading volume; a single trip with one driver and two collectors could cover ~ 1800 households. With a maximum of three trips per day a tipper truck with one driver and two collectors could cover ~ 5400 households. Similar to an auto-rickshaw tipper trucks can directly transport the collected waste to treatment centres without the requirement of additional secondary collection points and vehicles whenever the transfer distance is small.

To compares the number of different collection vehicles and their respective capex and opex required to collect the waste from 5000 households in a day assuming the average waste generation per household at 2 kg and the waste density at 350 kg/m^3. Further, Fig. 6.2 attempts to estimate and combine the collection fleet required for 10,000 households with scenario 1 as 50% of households having narrow road connections and 50% of households having wide road connections, and scenario 2 as 25% of households having narrow road connection and 75% households having wide road connection. The comparison in Fig. 6.3 makes it clear that the utilization of handcarts and tricycle has significantly lower capex costs in comparison with auto-rickshaws and tipper trucks. The Opex cost for handcarts and tricycles is considerably higher considering higher labour requirements.

The comparison of per year opex costs of different collection fleets are depicted in Fig. 6.3. For attending a lesser number of households utilization of handcarts and tricycle seems financially feasible however, as the coverage area increases the opex costs of handcarts and tricycle considerably increase in comparison with auto-rickshaws and tipper trucks. An increase in the opex cost for handcarts and tricycles is directly linked to the increasing number of labour requirements. For higher cost-effectiveness in waste collection and transportation, proper planning in the selection of the collection fleet is crucial. The utilization of handcarts and tricycle should be strictly limited to areas and areas with narrow roads. Whenever the possibility raises it is advisable to change the collection fleet from handcarts and tricycles to auto-rickshaws. Table 6.2 shows the number of vehicles needed in Senegal.

6.2.2.1 Transport Fleet Distribution with Respect to Region and Road Types of Senegal

The number of different collection vehicles that are necessary to collect waste from specific regions of Senegal is compared in Figs. 6.4 and 6.5. An attempt is made in Fig. 6.4 to estimate and combine the collecting fleet that is required for each region, taking into consideration that fifty percent of households have limited road

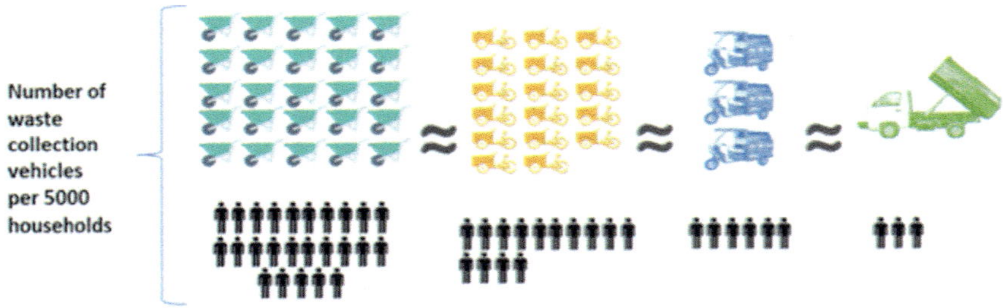

Parameter	Handcart	Tricycles	Auto rickshaw	Tipper truck
Collection frequency & model	Daily, Door to door	Daily, Door to door	Daily, door-to-door or block/kerbside collection	Block/Kerbside collection
Labour / vehicle	1 person	1 rider and 1 collector	1 driver and 1 collector	1 driver and 2 collector
Coverage per trip	~90 HHs	~180 HHs	~700 HHs	~1800 HHs
Collection time per vehicle trip	2.0 - 3.0 hours	3.5 hours	2.0 hours	2.0 hours
Max. trips per day/vehicle and total coverage	2 trips and ~180 households	2 trips and ~360 households	3 trips and ~2000 households	3 trips and ~5400 households
Capex* for attending 5000 HHs.	0.75 M CFA	1.40 M CFA	9.00 M CFA	11.00 M CFA
Opex/year** for attending 5000 HHs.	2.075 M CFA	1.26 M CFA	1.38 M CFA	1.34 M CFA

Combination of collection fleet required for 10000 households with 50% household having narrow road connection and 50 % having wide road connection

Capex*	14.48 M CFA
Opex/year**	2.57 M CFA

Combination of collection fleet required for 10000 households with 25% household having narrow road connection and 75% having wide road connection

Capex*	17.16 M CFA
Opex/year**	2.52 M CFA

*Capex is calculated by considering costs of handcart as CFA 30,000.00/unit, tricycle at CFA 100,000.00/unit, auto rickshaw at CFA 3,000,000.00 lac/unit, and tipper truck at CFA 11,000,000.00/unit.

** Opex is calculated considering common labour cost at CFA 80,000/month and 10% of capex maintenance cost for handcart and tricycle, and 20% of capex maintenance cost for auto rickshaw and tipper truck

Fig. 6.2 Overview of collection vehicles and their respective capex and opex costs

connections and fifty percent of households have large road connections. Taking into account that 25% of houses have limited road connections and 75% of households have wide road connections, Fig. 6.5 makes an attempt to estimate and combine the collecting fleet that is required for each region. The employment of handcarts and tricycles appears to be financially feasible for the purpose of catering to a smaller number of households; nevertheless, as the coverage area expands the operational expenses of handcarts and tricycles significantly increase in comparison to those of auto-rickshaws and tipper trucks. The amount of labour that is required to operate

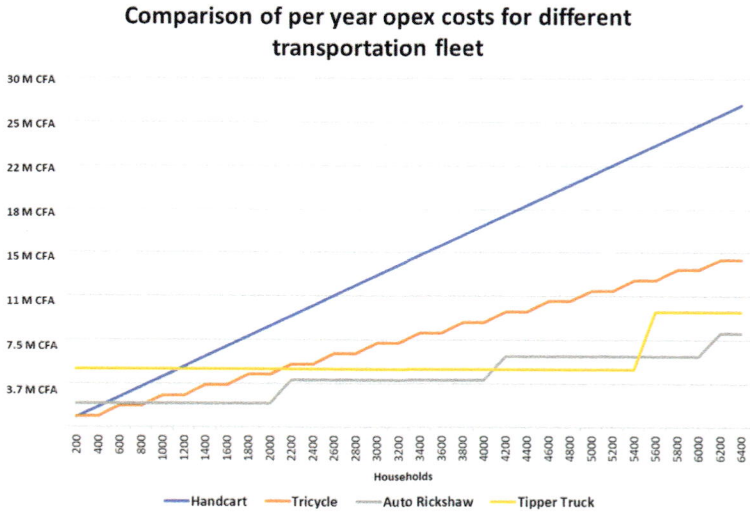

Fig. 6.3 Comparison of per year opex costs for different transportation fleet

handcarts and tricycles is directly proportional to the increase in the operational expenditures for these vehicle types. When it comes to waste collection and transportation, thorough planning in the selection of the collection fleet is absolutely necessary in order to achieve increased cost-effectiveness. The utilization of handcarts and tricycles has to be mainly restricted to areas and regions that have roadways that are small on design. It is recommended that the collection fleet be switched from handcarts and tricycles to auto-rickshaws whenever an opportunity is available.

6.3 Plan for Implementing Waste Storage and Closing of Collection Gaps

Waste collection fee

The economic viability of the collection system depends on user fees. Different schemes may be adapted to collect the fee from households based on the social conditions. The introduction of a user fee and collection in periodic patterns e.g. monthly or integrating waste collection fees with the property tax or utility are some solutions. To promote source segregation, different user charges for mixed waste disposal at the household level and additional environmental tax for commercial units could be considered in the long term.[8]

An official person e.g. waste collector should collect the user fees. To build up trust among citizens, the waste collector should hand out receipts for each received

[8] Financial aspect is explained detailedly in Chap. 10.

Table 6.2 Overview of collection vehicles needed in 14 regions of Senegal

		Average household size in Senegal as of 2021 is 10.8	Single per day trip can visit 90 HHs	Single trip cover ~ 180 households	Single trip with cover ~ 1800 households	Single trip of a compactor truck of 12 m^3 covers ~ 2000 households
Region	Population in 2022 by region	Number of household at 2022[a]	Number of handcart per day trip of collection	Number of tricycles per day trip of collection	Number of Auto Tripper trucks per day trip of collection	Number of compactor trucks per day trip of collection
Dakar	4,042,225.00	374,280.09	4158.67	2079.33	207.93	187.14
Thies	2,280,458.00	211,153.52	2346.15	1173.08	117.31	105.58
Kaolack	1,267,014.00	117,316.11	1303.51	651.76	65.18	58.66
Diourbel	1,980,821.00	183,409.35	2037.88	1018.94	101.89	91.70
Fatick	965,912.00	89,436.30	993.74	496.87	49.69	44.72
Kaffrine	782,273.00	72,432.69	804.81	402.40	40.24	36.22
St. Louis	1,150,062.00	106,487.22	1183.19	591.60	59.16	53.24
Louga	1,121,619.00	103,853.61	1153.93	576.96	57.70	51.93
Matam	789,219.00	73,075.83	811.95	405.98	40.60	36.54
Tambacounda	937,182.00	86,776.11	964.18	482.09	48.21	43.39
Kédougou	203,723.00	18,863.24	209.59	104.80	10.48	9.43
Kolda	875,660.00	81,079.63	900.88	450.44	45.04	40.54
Ziguinchor	729,951.00	67,588.06	750.98	375.49	37.55	33.79
Sédhiou	612,649.00	56,726.76	630.30	315.15	31.51	28.36

[a] https://www.statista.com/statistics/1228286/average-household-size-in-africa-by-country/

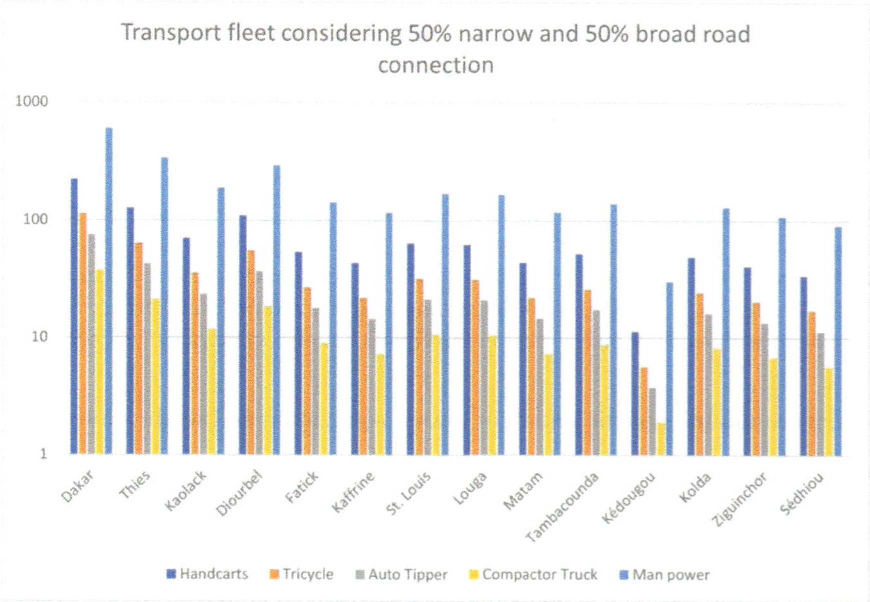

Fig. 6.4 Combination of transportation fleet required considering 50% household with narrow road connection and 50% household having broad road connection

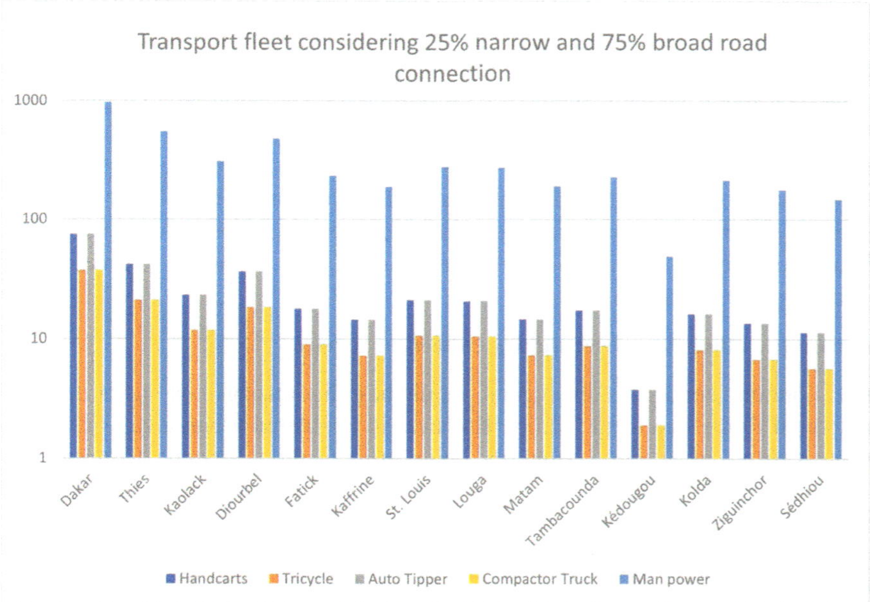

Fig. 6.5 Combination of transportation fleet required considering 25% household with narrow road connection and 75% household having broad road connection

fee. Further information on the financial system (received fees, operation and maintenance costs, etc.) should be reported annually in the form of an information campaign to ensure the system's liability and transparency. One option to enable greater transparency would be to implement double-entry digital accounting. As the name suggests double-entry bookkeeping relies on a two-sided accounting entry to maintain financial information. There are two equal and corresponding sides of the double-entry bookkeeping system: debits and credits. Double-entry bookkeeping is characterized by transactions affecting at least two accounts which always contain one debit and one credit and which have equal total debits and credits. The purpose of double-entry bookkeeping is to detect financial irregularities and fraud. Such a mechanism will not only builds trust in the citizen but also increases the acceptance level.

Promote community participation in source segregation and wastes collection

Participation of the community is crucial for a successful collection and treatment program. The citizens should be contacted through several channels such as community leaders, face-to-face interaction, awareness campaigns, advertising tools, and integration of awareness-raising directly to the waste collection vehicle and provide them with loudspeakers for creating regular awareness on complete waste management. The importance of source separation should be highlighted during introduction of the new collection scheme and the composting facility. Explaining how the community can benefit from waste management as a source of income, keeping a clean neighbourhood, and protecting public health would encourage the users to pay the user fee. In this regard the responsibility of both service providers and citizens should be clearly defined. With an effective community mobilization problems such as illegal dumping could be solved. This requires a stepwise program to first identify the vulnerable points, clean the area and allocate community storage bins to these points. Secondly raise people's awareness and introduce a traffic light system (red yellow green) for controlling waste disposal in the community bins. The target group should be informed regularly and in case of a deteriorating situation citizens should get an alarm (yellow notification for moderate and red for significant problems) for monetary penalties (see Fig. 6.6).

Increase the efficiency of secondary collection service

Under optimum conditions, the organic should be either managed on-site or directly transported to the treatment facility. However in populated urban areas with long distances between collection points and composting facilities, organic waste storage in the secondary collection points is inevitable. The most important issue in secondary points is the separate storage of organic waste. Galvanized iron mobile garbage bins should be placed for hygiene storage. The mobile compactors are suitable storage containers for areas with high waste generation rates. The waste should be weighed and transported on daily basis to the nearest treatment facility (Fig. 6.7).

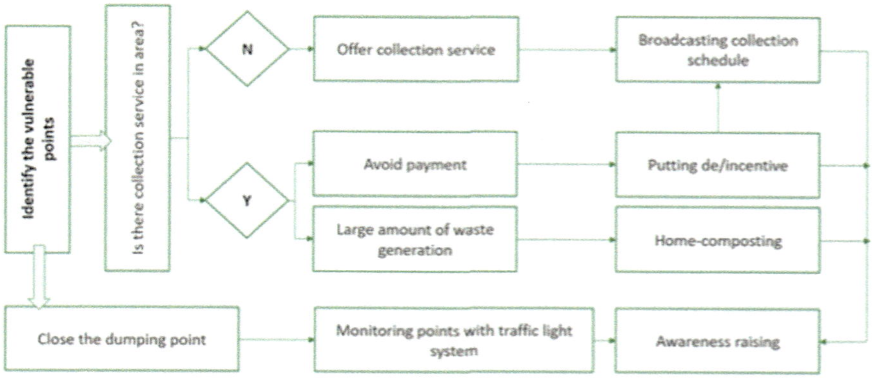

Fig. 6.6 Approach to managing the waste vulnerable points

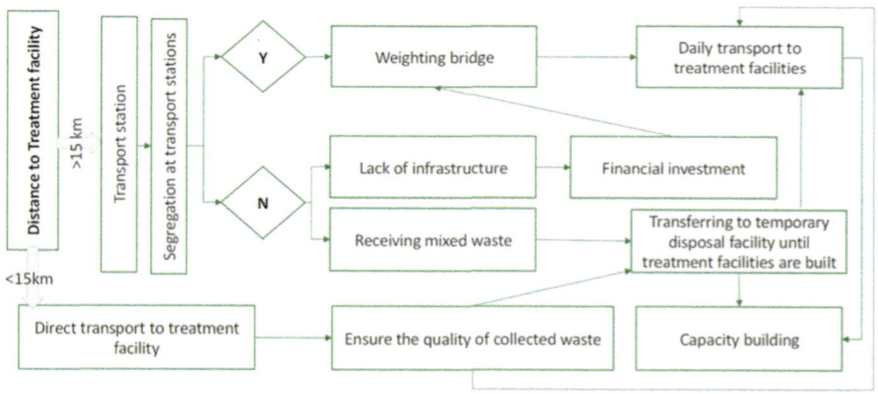

Fig. 6.7 Troubleshooting of secondary waste collection points

The collection service operators working on the secondary collection points and transfer stations should be trained by sanitary inspector to segregate and transport biowaste. The secondary collection points and transfer stations should be also monitored using the traffic light system. The workers should, citizens should get an alarm in case of a deteriorating situation (yellow notification for moderate and red for significant problems). They should be informed about the importance of separate storage for a value-added final product by visiting the composting facility. Overall in municipalities it is important to ensure the cost recovery of the collection services through effective collection of user fees. The modification of system design including number and type of collection vehicles, capacity building of waste workers and awareness raising would contributes to management of vulnerable points and separate handling of organic waste at secondary collection points. Accordingly the sustainability of the collection services and coverage is assured.

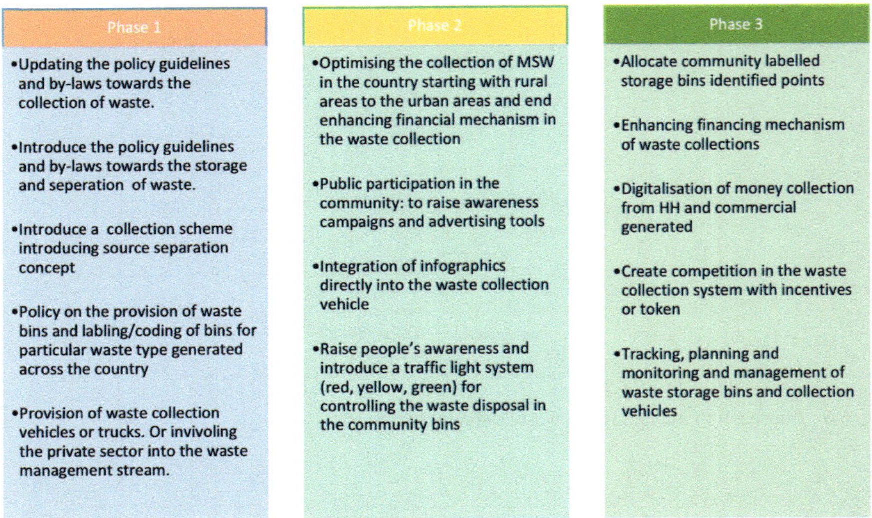

Fig. 6.8 Plan for implementing waste storage and closing of collection gaps

Following is the approach for promoting alternative MSW management techniques in Senegal that have to be divided into short-, medium- and long-term goals. The primary approach for achieving increment in the storage and collection of MSW in communities in Senegal has to do with updating the policy guidelines and by-laws to areas the collection of waste. Further optimising the collection of MSW in the country starting with rural areas to the urban areas and end enhancing financial mechanism in the waste collection as demonstrated in Fig. 6.8.

By implementing an optimized waste collection and transportation system it is possible for Senegal to achieve 100% segregated waste collection. This would have several significant benefits such as reducing the number of vulnerable points for municipal solid waste, increasing the value of organic waste by reusing it as animal feed, and allowing for the creation of centers to treat specific waste streams. Additionally it would increase the quality of compost by reducing impurities, generate more revenue for municipalities, create sustainable jobs throughout the waste management chain and reduce greenhouse gas emission.

References

Cabral et al. (2012) Low-level environmental exposure to lead and renal adverse effects: a cross-sectional study in the population of children bordering the Mbeubeuss landfill near Dakar, Senegal. Human & Experimental Toxicology

Cheikh Dieng BM (2020) Biomedical waste management in Dakar, Senegal: legal framework, health and environment issues; policy and program options. Cities & Health

Kaza S, Yao L, Bhada-Tata P, Van Woerden F (2018) What a waste 2.0: a global snapshot of solid waste management to 2050. Urban Development Series. World Bank, Washington, DC
Netherlands Enterprise Agency (RVO) (2022) Sector report circular economy Senegal. Terminomic B.V., Rotterdam
Silpa Kaza LY-T (2018) What a waste 2.0 a global snapshot of solid waste management to 2050. Wash
UCG (2016) Rapport campagne nationale de caractérisation des ordures ménagères et assimilés

Chapter 7
Strategy for Centralized Waste Treatment Facilities

7.1 Current Status of MSW Treatment Practice

Official landfills in cities that are well-equipped might supplement this with on-site works and the compaction of waste by machines as it accumulates when waste is piled up. The majority of dumpsite in Senegal however is operated on an informal basis. This practice is responsible for an enormous amount of greenhouse gas (GHG) emissions[1] and poses a risk to the local communities as well as the environment as it contaminates the soil, air and groundwater. Further the materials across the entire Senegal are rarely recovered from landfills demonstrating the prevalent linear consumption patterns. This is also caused by a lacking selective sorting system. In addition there is a limited amount of material recovery (non-professional manner) from landfills which is another evidence of the common linear consumption patterns. Inadequate implementation of a selective sorting mechanism is also to blame for this.[2,3] The informal recovery of materials such as iron, aluminium, plastics or food scraps is often accompanied by unsanitary and unhygienic working conditions. These conditions can be hazardous to one's health. The majority of the time waste pickers do their jobs without any sort of protection putting them in danger of sustaining injuries and developing respiratory or skin diseases.

Informal recovery efforts at the Mbeubeuss dump have developed into a functioning economy in recent years. Even though these examples show that waste recovery activities do exist the systems involved are still in their infant stage and lack a controlled and enforced management of externalities. Despite this Senegal has begun to implement high-value recycling practices for the majority of its resources. These include rejected metal plastic and paper products as well as organic waste electrical and electronic waste. In Senegal currently there are no options for the valorisation/

[1] Waste Wise Cities Bulletin (2020).

[2] Sector Brief Senegal: Solid Waste Management and Recycling. GIZ, 2021.

[3] Circular Economy in Africa-EU Cooperation (2020).

S. Narra et al., *Sustainable Waste Management in Senegal*, World Sustainability Series,
https://doi.org/10.1007/978-3-031-98968-1_7

recycling of glass. Waste picking is a prevalent activity in Senegal where waste-pickers gather at dumpsites across the country. Unfortunately, these workers are not under the protection or regulatory framework of the local or national government. This leaves them with little legal or physical security in an economy that is largely invisible. As a result waste-pickers are among the most susceptible and marginalized groups in Senegal. The work they do is hazardous and physically demanding with significant environmental and social risks associated with poorly managed dumpsites.

Tremex, Recuver, Refemetal, and Cosem SA have been processing thousands of tons of **metal** per month for over 15 years, mostly for sale to Spain. However, Société Métallurgique d'Afrique (Someta) is currently the main player in the industry and holds the top rank. With an annual recycling capacity of 50,000 tons of metal, SOMETA is a major player in the metal processing business in Africa, creating concrete reinforcing bars from waste materials. The company generates a turnover of XOF 5 billion (equivalent to EUR 7.6 million). SELMEG has a strong foothold in the market recycling approximately 6 tonnes of aluminium every year (comparatively low amounts). The metal recycling industry recycles up to six tons of aluminium annually according to the German Company for International Cooperation (GIZ). SinaPro based in Dakar specializes in gathering and selling scrap iron non-iron and computer scrap (E-Waste metal housings).

Plastic production plants play a vital role in creating plastic flakes and granules. Few processors use recycled plastic due to high investment but the demand for it is twice as that of the current supply. The market for recycled plastic shows strong growth potential. ProPlast is the leading plastic recycling company in Senegal renowned for their exceptional collection and processing of plastic waste into high-quality recycled plastic. They intend to exceed their current monthly purchase of 100 tons of plastic waste through the acquisition of 20 "Récuplast" kiosks. Transtech Industries is a pioneer in producing exquisite furniture using recycled plastic which they source from various recycling companies including ProPlast. Senegal boasts of about 40 plastic processing firms primarily located in the Dakar region catering to the ever-rising demand for recycled plastic which currently stands at over 15,000 tons with a tremendous potential for expansion.

The repair and recovery of **waste electrical and electronic equipment (WEEE)** involves disassembly and the sale and recycling of recovered materials. Although the treatment of WEEE is still an informal activity it has become a major concern for the Economic Community of West African States making it an emerging market with significant opportunities for the establishment of infrastructure networks for WEEE collection disassembly recycling and reuse. The disposal of hazardous components that cannot be recycled or reused remains a real challenge. Gravita Senegal is a leading West African manufacturing recycling and processing unit for lead metal oxides and products specializing in non-ferrous recycling. The company manufactures re-melted lead and polypropylene chips from various sources including scrap from batteries battery plates and concentrates. According to the enterprise Gravita manufactures lead metal through smelting and recycling followed by refining alloying and manufacturing lead oxide.

The **paper and cardboard recycling industry** in Senegal has immense potential for growth. PRONAT operates as the sole plant in Senegal's paper and cardboard recycling industry utilizing a modest 2 tonnes of paper daily. This amount though is less than 1% of the national recyclable waste representing a mere fraction of the potential demand of 7%. Despite the current use of recycled paper to produce egg trays the recovery rate of organic matter paper and cardboard in Senegal remains unacceptably low. However the market for these materials is still in its early stages presenting a highly promising opportunity for growth and expansion. The dumping of waste in dumping sites is the waste treatment option that is utilized most predominantly in Senegal.

Senegal generates more than 20,000 tons of **medical waste** per year about 80% of the waste typically produced by health-care providers are coming mostly from the administrative kitchen and housekeeping functions at health-care facilities are non-hazardous and comparable to domestic waste.[4] However the rest of the wastes are about 15% of infectious wastes. Senegal has 50 sanitary districts (SDs) including 10 SDs in Dakar region; 12 major hospitals, 19 health centres, 122 dispensaries, 41 health units, 524 pharmacies and 692 private clinics.

In the **health sector** of Senegal the private sector[5] in collaboration with the Government offer a comprehensive treatment solution in 2022 that included sorting collecting and decontaminating infectious medical waste. A treatment centre is built and functioning to treat biomedical waste in Dakar. Other regions continue to dispose of biomedical waste in artisanal ovens or out-of-date models of incinerators releasing enormous clouds of toxic gases chlorinated organic particles and heavy metals into the air and posing risks of environmental degradation soil and water contamination and human and animal poisoning. The practice of open burning which is common in many healthcare facilities across the country is comparable to inefficient and improper incineration.

7.1.1 Drawbacks

- **No option for waste treatment at household**: nonexistence of any method to treat organic waste at household
- **No option for decentralised waste treatment option**: non-existence of any method to treat organic waste at decentralised facility and communities
- **No option for waste recovery and waste to energy centralised facility**: Non-existence of centralised waste treatment options throughout country
- **Absence of sanitary waste, hospitals and hazardous waste treatment options**: There is no option for treating sanitary waste from the households, hospitals and hazardous waste. A significantly high amount of the sanitary waste generated is currently being either dumped or burnt by households uncontrollably.

[4] World Health Organization (WHO) 2014.

[5] ECOSTERY 75+ (Dr. Sett).

7.2 Current State of Final Disposal of the MSW

Landfill remains the main means of disposal of collected waste in Senegal. Waste disposal involves two main local realities: in Dakar collected wastes are dump in Mbeubeuss landfill while in the rest of regions across the country and in households without collection services uncontrolled dumping and burning are the custom. In Senegal, over 2.9 million tonnes of waste are produced annually; of that amount, approximately 1.3 million tonnes are not collected. Of the waste that is collected, the majority is dumped at uncontrolled disposal sites located throughout the country, primarily Mbeubeuss, which is among the top ten largest disposal sites worldwide. Each vehicle collects about 14 tons of waste 334 disposals and estimate of 4676 tons of waste disposed of daily.[6]

The waste management issue plaguing Thies (Senegal's third largest city) is dire. Currently waste is being carelessly dumped in open areas without any proper waste recovery system in place. This unacceptable practice needs to be urgently addressed and changed in a step-by step manner to acceptable and sustainable practices. Six designated trucks transport waste to these unfenced dumpsites which are frequently scavenged by individuals looking for reusable or sellable objects (valuables). In addition domestic animals rummage through the waste and the wind carries the waste into neighbouring households. The community's well-being is at stake and immediate action must be taken.[7]

The polluter-pays principle mandates that industries and hospitals transport their waste to disposal sites including the hazardous waste that they generate. Strict safety measures must be followed during transportation whether it is done internally or externally. Proper storage of waste is a critical element in preventing toxic substances or pathogens from being dispersed. If the waste generator lacks disposal facilities specific companies authorized by the state must assist to handle hazardous waste.

Overview of the existing dumpsites and landfills

Senegal has no engineered sanitary landfill, but every region has dumpsites with no waste treatment before disposal. However in 2020 The Government of Senegal with the support of the World Bank implemented the project for the Promotion of Integrated Management and Economics of Solid Waste (PROMOGED); a budget of €259 million is set aside for the rehabilitation of existing dumpsites across regions in Senegal. PROMOGED project selected dumpsite includes the following[8]:

- Mbeubeuss landfill in the Dakar region;
- Gandon, Richard Toll, Rosso, Goureg Serigne dumpsites located in the Northern pole of Senegal;
- Thiès, Gandigal, Tivaouane dumpsites located in the Thiès pole;

[6] Beri (2018).

[7] Circular Economy in Africa-EU Cooperation (2020).

[8] https://www.promoged.sn/en/en/main-lines-action.

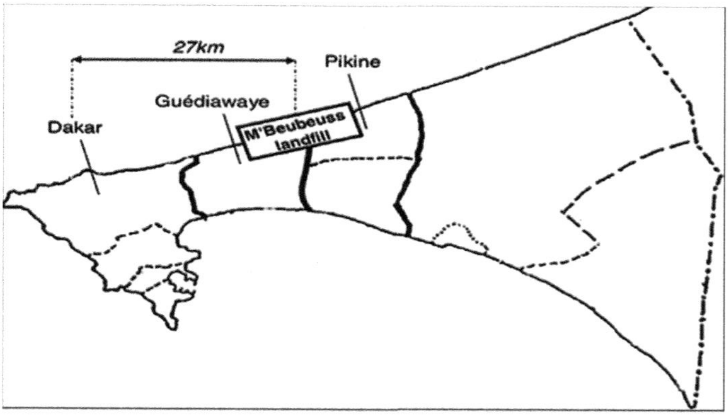

Fig. 7.1 Map of showing the distance from the capital city Dakar to the Mbeubeuss landfill[11]

- Route Diendé, South Airport, Mamatoro dumpsites located in the Casamance pole.

The PROMOGED project will establish three technical landfill sites at Touba, Tivaouane and Kaolack.

Mbeubeuss disposal site is one of the largest open landfills in Africa and is located outskirts of Dakar about 27–30 km from the city centre as shown in Fig. 7.1. It is the main solid waste disposal site for the capital city of Dakar with a projected population of approximately 4 million people as of 2022. The famous Mbeubeuss was created in 1968 of a size of 14 ha on a drying lake and sits on a flood plain close to the sea. Moreover as of 2017 the site had expanded to 114 ha from 14 ha in 1978 and continues to expand due to a rapid increase in population and urbanization. About 1300 tons of waste are brought in each day by 230 trucks estimating a total of 475,000 tons of waste dumped each year.[9] State-trained control agents measure the amount of waste entering the Mbeubeuss landfill using a weighbridge. An entire community of about 4000[10] or more informal recyclers or 'waste pickers' has been surviving a living at the Mbeubeuss dumpsite since 1968.

The informal recyclers or waste pickers sector plays a large role in waste management activities especially on the landfill; immediately a dump truck arrives pickers do more than scavenge; based on their specialty and interest they choose plastic, cans, wood, metal, clothes and even organic waste sold as fodder to animal farms. Most of the time the waste pickers pick through until nothing is left.[12]

---

[9] Netherlands Enterprise Agency (RVO) (2022).

[10] IDRC (2011).

[11] Cabral et al. (2012).

[12] Circular Economy in Africa-EU Cooperation (2020).

7.2.1 Drawbacks

Open burning of waste: incineration of waste occasionally to create space for more waste leads to the emission of greenhouse gases and acid rain from the interaction of NOx, sulphur dioxide, chlorides, fluorides and rainwater has an adverse effect on agriculture.

Uncontrolled and unfenced dumpsites: Senegal has more than 40[13] uncontrolled and unfenced dumpsites that are harmful to the environment. People who hope to get objects that can be reused or sold exploit unfenced dumpsites and are being exposed to harmful substances. Domestic animals loiter around dumps in search for food. Wind disperses waste into neighbouring households.

Unnecessary small dumpsites: Settlements create small and uncontrolled dumpsites due to improper waste collection systems and no availability of landfills. The closure and reclamation of these dumpsites can be done with greater ease that could not only result in the reduction of waste footprint but has the potential to make more land space available for SWM and other additional activities.

Absence of a buffer zone between human settlements and dumpsites: The provision to buffer zones around the disposal facility is mandated in the MSW Rules. Due to an entire community of informal recyclers or 'waste pickers' surviving a living next to dumping sites. About 4000 settlements of informal recyclers or 'waste pickers' are living at the Mbeubeuss dumpsite since 1968.

7.3 MSW Treatment Hierarchy

It's important to note that the waste management hierarchy promotes a holistic approach to waste management emphasizing waste reduction and resource conservation. While recycling is generally considered more preferable than energy recovery due to its potential for resource conservation the specific preference may depend on the waste composition available infrastructure, environmental considerations, and regional policies. The hierarchy encourages a combination of approaches to achieve the best overall waste management outcomes based on the principles of waste reduction, reuse, recycling and energy recovery. According to the waste management hierarchy waste prevention and reduction are the most preferred waste management methods, followed by reuse recycling and energy recovery (see Fig. 7.2). The waste management hierarchy is a prioritization framework that guides waste management decisions based on their environmental impact. The specific order of preference may vary slightly depending on regional or national waste management policies but in general the hierarchy is as follows:

[13] https://www.giz.de/en/downloads/SectorBrief_Senegal_Waste.pdf.

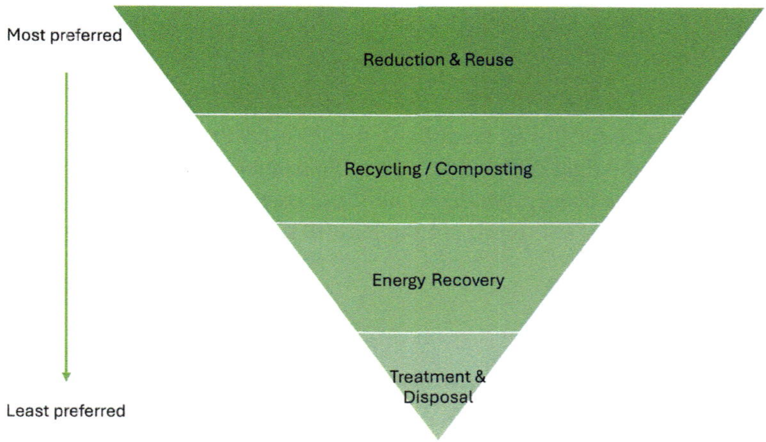

Fig. 7.2 Waste management hierarchy according to EPA

Waste Prevention: The most preferred method is to prevent or minimize the generation of waste in the first place. This involves reducing waste at the source through practices such as product design, packaging reduction and encouraging sustainable consumption habits. Waste prevention or minimisation strategies have been discussed in previous sections of this document.

Reuse: Reusing items is the next preferred option. It involves extending the life of products by repairing refurbishing or repurposing them for their original or alternative uses. Reuse helps conserve resources and reduce the need for new product manufacturing.

Recycling: Recycling is the third preference in the waste management hierarchy. Recycling is the process of converting waste materials into new products reducing the need for raw materials and minimizing environmental impacts. It typically involves sorting, processing and reprocessing materials like paper plastics metals and glass to manufacture new products. Recycling promotes resource conservation reduces landfill waste and helps mitigate pollution associated with extracting and processing virgin materials. It is often favored for materials with well-established recycling markets such as paper and metals where the infrastructure for collection, sorting and processing is well developed.

Energy Recovery: Energy recovery/RDF production, which includes waste-to-energy processes like incineration or gasification is the next step in the hierarchy. RDF is a processed form of non-recyclable waste that is used as a fuel source in waste-to-energy facilities. It involves sorting and processing residual waste to remove contaminants and increase its energy content. RDF production offers an alternative to landfilling non-recyclable waste, helping to reduce the volume of waste going to landfills and generating energy through combustion or gasification. RDF can be used as a substitute for fossil fuels in cement kilns, power plants, or industrial boilers.

RDF production is typically favored when the waste stream contains a significant proportion of non-recyclable materials and when waste-to-energy infrastructure is in place.

Disposal: Disposal typically in landfills is the least preferred waste management option. It involves the final disposal of waste that cannot be prevented reused recycled or recovered for energy. Modern landfill designs aim to minimize environmental impacts but they still represent the least desirable option in terms of waste management hierarchy.

7.3.1 Waste to Energy Treatments Facilities and Landfill Suitability Assessment

When selecting a suitable waste treatment technology it is important to consider the amount and type of waste generated as well as sustainability factors. This report offers a more detailed overview of the quantity and composition of municipal solid waste (MSW) in various cities and regions in Senegal. The study gathered primary data on waste quality and quantity through desktop research and made projections based on calculations. Figure 7.3 shows each stream has significant potential because of its diversity in types of solid waste and demonstrate the alternative waste treatment options for each waste composition streams generated in the country. The potential rate of organic recovery from putrescible waste is 12%. Twenty five percent of the resource is expected to be recovered mainly iron and various metals, plastics, paper or cardboard. However 63% of waste is considered to be unusable on the other hand but within the presumed non-usable composition of 53% of sand can be recovered and reused. Another concept can be involved where the waste incineration technology can hold more than 40%. This share can be supplied from the streamline of the landfill, excluded fractions from composting and MRF facilities and RDF as can be seen in the Fig. 7.3. Furthermore, waste incineration has a vital issue as a big advantage point that it accommodates mixed waste without any pre-treatment procurement.

The criteria and sub-criteria used to choose the best waste treatment technology are technical, economic, social and environmental considerations. In Table 7.1 these requirements are concisely discussed.

7.3.2 Data Accusation and Processing

Based on the data available at the countries level in Senegal gathered data are divided into seven criteria regarding the socio-cultural, environmental and technical/economic factors. Table 7.2 illustrates those criteria with their sources where the power line criterion is assumed from the main road network dataset (primary, secondary, and tertiary roads) of the country.

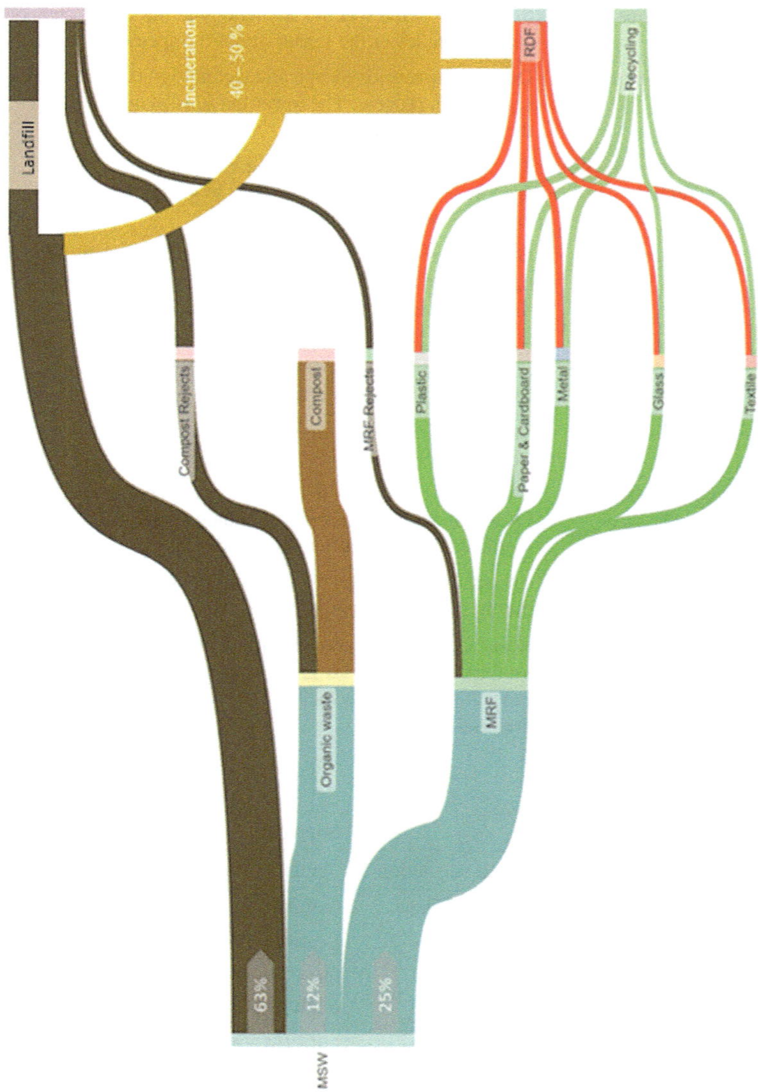

Fig. 7.3 Treatment options for Municipal solid waste generated

Table 7.1 Criteria selected for the evaluation of the waste treatment technology (Alam et al. 2022)

Criteria	Sub-Criteria	Description
Technology	Waste quality and quantity	How much a specific technology is compatible with the composition, calorific value and quantity of the available waste from the city
	Continuity of supply	Sensitivity of the specific technology to the consistency of supply. The lower the sensitivity of the specific technology
	Waste Segregation	Define how sensitive a specific technology is to the waste segregation process. The lower the sensitivity, the better the technology is
	Energy generation potential	The amount of electricity produces from waste using a specific technique. It is preferable to generate as much energy as possible from waste
Environment	GHG emission	GHG emission Greenhouse gas may be emitted during the operation of a specific technology. The lesser the emission, the better the technology
	Production of hazardous waste	Various hazardous wastes like dioxin, furan, etc. can be produced during the operation of a specific technology. The lesser the emission, the better the technology
Economy	Capital cost	This is the initial investment needed for the plant to start. Low-cost technology is recommended
	O&M cost	The amount of money that will be spent on the plant's Operation and maintenance over the course of its life. It's worth should be kept to a minimum
Social	Safety of public health	The specific technology should not impose health hazards during the plant's operation
	Social acceptance	Resistance by local people can intensely affect the establishment of waste-to-energy plants inside an area. Innovation with high social acknowledgment is generally preferred

Table 7.2 Data accusation

Data		Sources
Environmental	Waterbody/flooded vegetation	Sentinel-2 Land Use Land Cover
	Forest/agriculture/rangeland	Sentinel-2 Land Use Land Cover
Socio-cultural	Settlement	Sentinel-2 Land Use Land Cover
	Sensitive area	https://www.openstreetmap.org and Sentinel-2 Land Use Land Cover
Technical/economic	Road	https://www.openstreetmap.org
	Slope	https://earthexplorer.usgs.gov/
	Powerline	https://www.openstreetmap.org

To determine the best locations for waste treatment plants and landfills this work uses two methods of processing namely: restriction map and suitability map processing. The data was prepared, analyzed with various criteria's and have been standardized through the fuzzification method. The work also assigned weightages to each criteria based on pairwise comparison matrix (PCM) with a consistency ratio of 0.093 (see Tables 7.3 and 7.4). This creates a suitability layer that helps in determination of the relative weightage of each set of criteria. To exclude unsuitable areas buffer zones of 1000 m have been created to protect waterbodies/flooded vegetation, sensitive areas (such as bare land, natural reserves, parking, hospitals, and industrial zones) including slopes above 15%. This prevents the interference/degradation of natural conservational areas. These areas are assigned the binary value of "0" to indicate they are not allowed. Finally the restriction layer and suitability layer have been combined using the fuzzy overlay operator "SUM" to obtain the final suitability layer. This final layer then identifies all possible suitable sites for waste treatment plants and landfills based on specific pre-set parameters on criteria standardization.

Table 7.3 Pairwise comparison matrix of criteria

Criteria	Settlement	Slope	Power	Road	Sensitive area	Waterbody/ flooded vegetation	Forest/ agriculture/ rangeland
Settlement	1	1/2	1	2	8	6	5
Slope	2	1	3	3	4	6	6
Power	1	1/3	1	1/4	3	6	3
Road	1/2	1/3	4	1	5	4	6
Sensitive area	1/8	1/4	1/3	1/5	1	1	1/3
Waterbody/ flooded vegetation	1/6	1/6	1/6	1/4	1	1	2
Forest/ agriculture/ rangeland	1/5	1/6	1/3	1/6	3	1/2	1

Table 7.4 Criteria weightage

Criteria	Weight
Settlement	0.222
Slope	0.309
Powerline	0.136
Road	0.199
Sensitive area	0.039
Waterbody/flooded vegetation	0.044
Forest/agriculture/rangeland	0.048

7.3.3 *Waste Treatment Plant and Landfill Site Suitability for Senegal Considering Various Different Parameters*

The maps (Figs. 7.4 and 7.5) display the most appropriate areas for placing a waste treatment facilities and landfill with blue indicating suitable areas and red indicating unsuitable areas where the coastal line data from Senegal was not considered in these maps as these areas are sensitive and should be excluded (given a suitability index value of "0"). The suitability areas were determined based on several criteria including slope settlement road and power line with their respective importance weights being 0.309, 0.309, 0.199, and 0.136. As a result both technical/economic and socio-cultural factors played an important role in the siting process in this study. To ensure that the siting process was relevant and appropriate, the maps were generated using different scenarios and buffer distances of 1000 m around sensitive areas and waterbodies. It's important to note that while the same criteria were used to evaluate the suitability of both the waste treatment plant and landfill the standardization method for distances was adjusted only for sensitive and settlement criteria as these are typically the most critical factors to consider when siting hazardous waste disposal facilities, landfills and treatment facilities.

The potential areas extracted per region from the scenario of 1000 m buffer resulted as the best one based on a continuous suitability index ranging from 0.60 to the highest values. This is due to the non-chaotic nature of the data used giving continuous

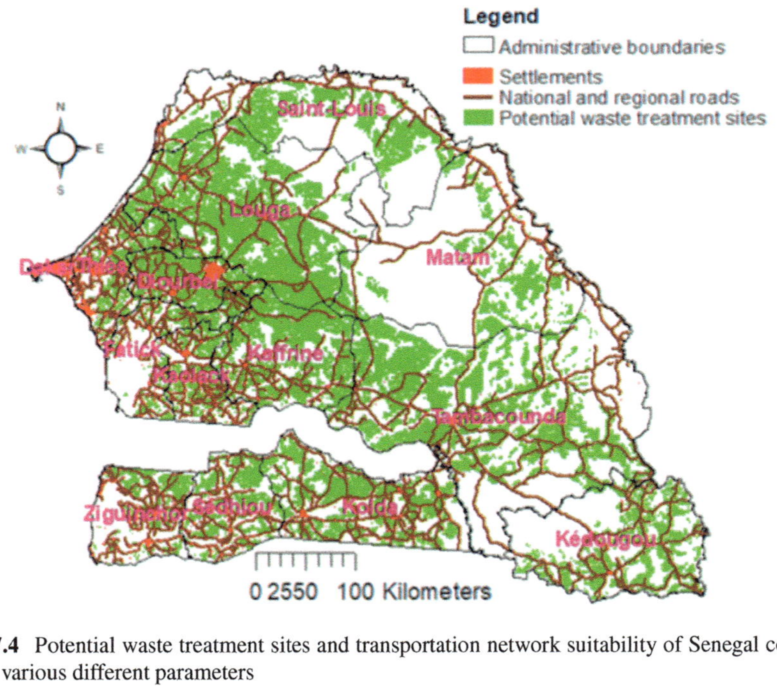

Fig. 7.4 Potential waste treatment sites and transportation network suitability of Senegal considering various different parameters

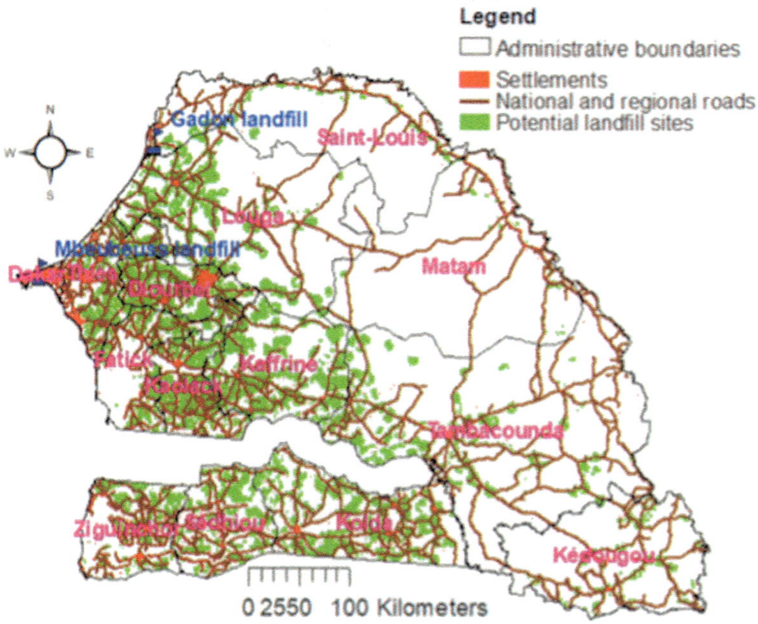

Fig. 7.5 Landfill sites and the transportation network suitability of Senegal considering various different parameters

information on land use and land cover in this assessment to get all possible potential sites for landfill and/or waste treatment plant implementation. Finally it would be possible to obtain definitive sites as a result of taking account of the waste disposal centres as hot spots for settlement and further optimization could take place by defining specific areas on the ground where location allocation techniques are used for commercial purposes. Further the proposal for landfill and waste treatment plants to be located at closed locations has been suggested in order to achieve a sustainably optimised treatment process through improved protection of workers and settlement as well as lower operational costs.

7.4 Centralized Facility to Manage Organic Waste (Composting)

7.4.1 Overview of the Size of the Region and Zone-Wise Centralized Composting Facility in Senegal

A centralized composting facility in Senegal can bring significant environmental and economic benefits by effectively managing the high-quality organic waste that currently lacks proper disposal methods. It is advisable to create a centralized composting facility per region and branding the end product per region. The decentral compost after reduction in its volume can be bought to the centralized facility to mature from many decentral composting plants collected together (increasing transport efficiency). The bagging and packaging can be done at the centralized facility and can be distributed in stores for sale.

Composting methods are manifold. Generally, the differences only concern the first phase of the composting process, i.e. the preliminary phase (see also chapter on the composting phase). The subsequent stabilization phase generally takes place in piles. Basically, a distinction can be made according to the handling or moving of the material, which can take place in batch (static methods, e.g. windrow composting, tunnel composting), semicontinuously (semi static methods, e.g. table composting) or continuously (dynamic methods, e.g. drum reactors). In the following, the most important methods for industrial scale composting will be discussed in Table 7.5.

Potential advantages of having a centralized composting facility per region:

Environmental Benefits

- **Waste Reduction**: The facility would enable the diversion of a substantial portion (12%) of total generated waste from landfills or incineration, reducing the strain on existing waste management systems and landfills.
- **Greenhouse Gas Emissions Reduction**: Composting organic waste instead of landfilling or incinerating helps prevent the release of methane a potent greenhouse gas. Methane emissions contribute to climate change, and composting reduces this environmental impact.
- **Soil Health Improvement**: The compost produced by the facility can be used as a nutrient-rich soil amendment, enhancing soil fertility, structure and water retention. This promotes sustainable agriculture and reduces the need for synthetic fertilizers, which can have negative environmental consequences.
- **Water Quality Protection**: By diverting organic waste from landfills, the facility reduces the risk of leachate contamination, which can negatively impact groundwater and nearby water bodies.

Table 7.5 Type of centralised composting technique, their benefits, and drawback

TYPE	Benefits and drawbacks	Recommendation	Model
Windrow composting	**Benefits**: low investment costs and a low level of process control **Drawbacks**: More space needed	The waste is piled in triangular or rectangular heaps, which are turned to enhance aeration and homogenization The need of turning can be reduced by installing air piping under the pile, providing passive aeration (chimney effect) or aeration by over- or under pressure. The windrows can be placed in open air or in a closed hall	  *Source* https://www.biobasedeconomy.eu/app/uploads/sites/2/2017/07/centralized-composting.pdf
Table composting or Row composting	**Benefits**: Allow a semicontinuous composting process. Less space, the aeration conditions are more homogeneous and aeration and watering can be adjusted according to the decomposition phase **Drawbacks**: machinery investment needed	Table composting takes place in a hermetically sealed composting hall, in which the organic waste is piled up into a stack and constantly aerated through perforated floor panels. A fully automatic turning machine moves the organic matter from one end where new shredded organic matter is added to the other end where the finished compost can be removed	 A fully automated turning machine *Source* Nojosa et al. (2021) 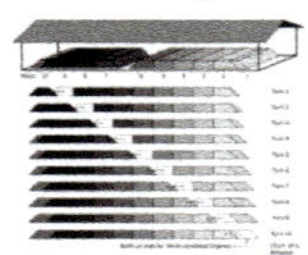

(continued)

Table 7.5 (continued)

TYPE	Benefits and drawbacks	Recommendation	Model
In-Vessel (Tunnel or Drum reactors) composting	**Benefits**: andle large volumes of any type of organic material and can operate in any climate year around Allows for a precise monitoring of process parameters and control of emissions **Drawbacks**: High capital cost. A higher degree of expertise and training in operating in-vessel systems is required	An "in-vessel" compost option involves mixing and making compost in a fully "environmentally controlled" enclosed operation, such as a drum or container with an agitating mechanism Tunnel composting units are large-scale rectangular vessels employing forced aeration systems Preliminary decomposition takes place in closed tunnels or containers (also called biocell reactors)	 *Source* https://www.eggersmann-recyclingtechnology.com/en/recycling-plants/biological-treatment/tunnel-composting/ 

Economic Benefits

- **Job Creation**: The establishment and operation of a centralized composting facility would create employment opportunities in various sectors, such as waste management, compost production, facility maintenance, and transportation.
- **Resource Recovery**: Composting organic waste allows for the recovery of valuable nutrients, such as nitrogen, phosphorus, and potassium, which can be used as fertilizer for agriculture. This reduces the reliance on imported chemical fertilizers resulting in cost savings for farmers and the agricultural sector as a whole.
- **Revenue Generation**: The facility can generate revenue through the sale of compost and related products, such as organic fertilizers, soil amendments and mulch. This revenue stream can contribute to the financial sustainability of the facility.
- **Waste Management Cost Reduction**: Proper management of organic waste at the source, through composting, can potentially reduce the costs associated with waste collection, transportation, and landfill disposal. This cost reduction can benefit both municipal authorities and waste generators.

Overall the creation of a centralized composting facility in Senegal would lead to a more sustainable waste management system, with positive environmental impacts such as waste reduction, greenhouse gas emissions reduction, improved soil health and water quality protection. The facility would bring economic benefits through job creation, resource recovery, revenue generation and waste management cost reduction.

7.4.2 *Overview of Region or Zone Wise Centralized Composting Facility in Senegal*

The overview of the area required to set up centralised composting plant (in ha) and cost required to set up a centralized composting facility in all the 14 regions of Senegal (see Table 7.6).

Table 7.6 Overview of the size of the centralized-composting facility in Senegal

Zone	Region	Population in 2050 by region	Approximate area required to set up centralised composting plant (in ha)[a, b]	Approximate area required to set up centralised composting plant in (in Billion CFA)[c]
Dakar	Dakar	8,978,949.95	30.34	8.65
Cap Vert-Thies	Thies	5,065,556.28	8.32	2.08
Central Senegal	Kaolack	2,814,404.27	4.62	1.16
	Diourbel	4,399,975.91	7.22	1.81
	Fatick	2,145,569.71	3.52	0.88
	Kaffrine	1,737,654.41	2.85	0.71
Total area in acers			18.22	4.55
Northern Senegal	St. Louis	2,554,620.08	4.19	1.05
	Louga	2,491,439.95	4.09	1.02
	Matam	1,753,083.49	2.88	0.72
Total area in acers			11.16	2.79
Tambacounda	Tambacounda	2,081,752.07	3.42	0.85
	Kédougou	452,527.66	0.74	0.19
Total area in acers			4.16	1.04
Casamance	Kolda	1,945,093.93	3.19	0.80
	Ziguinchor	1,621,432.13	2.66	0.67
	Sédhiou	1,360,870.49	2.23	0.56
Total area in acers			8.09	2.02

[a] Nojosa et al. (2021)
[b] Size of post-composting unit considering 60% biomass weight reduction after composting at decentralized and centralised composting facilities, and transfer efficiency from decentralized composting facilities to the centralised post-composting facility as 80%
This is a preliminary theoretical estimation

Design consideration

The following factors are put into consideration in creating the concept of linking decentralized facilities to a centralized composting facility:

- **Site considerations**: The plant can be constructed near the landfill to ease the transportation of rejected materials. The design of the compost facility includes the consideration of the local environmental regulations for setting and designing waste processing facilities.
- **Climate**: rainfall, temperature, wind directions, etc. will affect the operation and efficiency of the windrow composting especially during rainy seasons. Based on the location's climatic conditions construction of the roof for the active and curing phases of the composting process should be decided.

The suggested concept covers the following parameters:

Raw feedstock: The characteristics of the organic materials that are received in the centralized facility and the output of the decentralized facilities are very important to the post-composting process. Both streams should have less than 10% impurities once they enter the curing phase. Therefore, source segregation, separate collection, and pre-treatment of organic waste are vital for a functional concept. It was assumed that the centralized plant has a capacity of 33 TPD (10,000 TPA) and that it receives 13 TPD of processed organic waste from the decentralized facilities within the region.

Sorting plant: A simple and cost-effective scheme is suggested for the sorting plant. The sorting line includes a bag opener, a conveyor belt with a hopper, two trommel sieves, and two magnetic separators. Waste bags are entered into the bag opener followed by the trommel sieves to separate the fine fraction. The over-sieve fraction contains recyclable and non-recyclable fractions, whereas the fine fraction mainly consists of organic materials. To ensure less than 10% impurities in the fine fraction, the foreign materials especially glasses should be removed manually from the under-sieve fraction.

Composting area: The land area required for windrows and movement of equipment is a crucial parameter. The suggested composting method is turning piles performed through a turner machine or a vehicle equipped with a mechanical shovel. The layout of composting area allows smooth and safe operations. The process of composting is undergone three phases: active, curing and finishing. For each phase the area necessary for material management operations was sized. The active phase of composting begins after the fine fraction of the sorting facility is piled and lasts 30 days. Afterward the piles are moved to the curing area. The turning has a lower frequency during the curing process. The pre-treated materials from decentralized facilities are also unloaded in windrows in the curing area. The green and wood waste can be used as structuring materials to lower the substrate moisture and to fix the C/N ratio in the unloaded piles. The piles remain on the ground for 30 days before sending to the sieving process. The stabilized compost is then stored for 60 days to become fully matured.

To design the windrow piles, the possible geometry of the windrow section (triangular, trapezoidal, and semi-circular) was examined and the option that takes the least land space was adapted as seen in Fig. 7.6. The volume of each pile was estimated by assuming that the density of the fine fraction of organic waste is about 800 kg/m^3. The stabilized organic waste and pre-treated materials received from decentralized facilities are assumed to have a density of 500 kg/m^3. The total area covered by the piles was calculated by knowing the total volume of waste inflow into the composting phase, the piles, section area and its length.

For calculating the composting areas, the following consideration was taken expressing the surface underlying the pile with base B (m), height H (m), and length L (m):

- Active area: $B = 5$ m, $H = 2$ m, L = 30 m
- Curing area: $B = 5$ m, $H = 2$ m, L = 50 m
- Finishing area: $B = 5$ m, $H = 4$ m, L = 50 m.

A volume contraction factor equal to 0.7 was employed to consider mass and volume reduction after the active phase of composting. After the curing phase this factor was considered 0.9. The width of the free perimeter between piles and the band around each pile, both along the sides and in the heads was considered equal to 4 and 5 m respectively to allow vehicle operations. The areas required for the different stages of the composting process are respectively 0.4 ha for the active phase 0.7 ha for the curing phase and 1 ha for the sieving and storage of the material. Table 7.7 depicts the results of the sizing procedure. The best pile shape for the active and curing phases is semi-circular and trapezoidal for the finishing phase.

Screening: A mature compost still contains coarse materials. The particle size of the finished materials depends on the physical properties of the feedstock and turning scheme. The sieving process is required to refine the final product. Rotating drum sieve is suggested for the compost plant. Based on the final application and local condition various mesh sizes can be used.

Leachate collection: The layout allows the collection of leachate generated. The piles in all composting areas are aligned parallel to the 2% slopes. In the active composting area, 2% lateral slopes are considered to ease the leachate flow. The surfacing is asphalt/concrete for efficient operation and leachate collection. The collected leachate is gravity-fed to the evaporation/drying lagoon for treatment. The

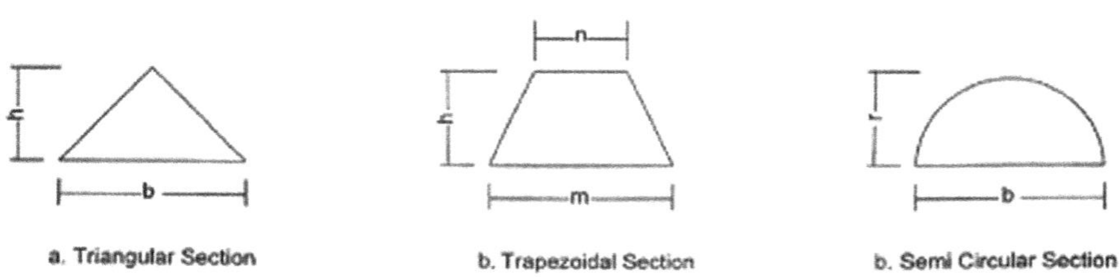

Fig. 7.6 Calculating the composting areas

Table 7.7 Plant sizing for 20 TPD fresh organic fractions and 13 TPD pretreated materials from a decentralized facility

Compost area sizing	Active	Curing		Finish	
		Centralized	Decentralized	Centralized	Decentralized
Number of windrow	3	1	2	1	2
Length of composting area (m)	47	39	42	64	67
width of composting area (m)	40	60	60	60	60
Area required (m^2)	1869	2361	2517	3858	4042
Total area (ha)	0.2	0.5		0.8	

evaporation lagoon is a large shallow pond or basin that is designed to allow the leachate to evaporate through exposure to the sun and air. As the leachate evaporates, the water vapor is released into the atmosphere while the concentrated nutrient-rich solids settle to the bottom of the lagoon. The solids can then be removed and disposed of according to regulatory requirements while the clean water vapor is released back into the environment. The evaporation lagoon is an cost-effective and sustainable method of managing the leachate produced during windrow composting, as it helps to prevent environmental contamination and reduce the volume of waste that needs to be disposed off.

Further rainwater harvesting can be part of the water management strategy. This method involves the use of the ground surface to collect rainwater. The collected rainwater is then directed to a storage tank or cistern, where it can be stored to provide water for irrigation, watering plants or other purposes. This can help to reduce the reliance on municipal water supplies conserve water resources and reduce the amount of storm water runoff.

Operation control: A weighbridge is used to accurately measure the weight of incoming and outgoing trucks carrying composting materials. This helps the facility to track the amount of materials that are being processed, as well as to ensure that trucks are not overloaded which could pose a safety hazard. Besides the incoming materials should be inspected for the rejection of hazardous materials.

Process monitoring: The concept includes measurement of temperature and moisture to establish a turning and irrigation schedule. Process monitoring involves the capacity building of workers to use sensors and other monitoring devices to continuously measure and record data on the conditions within the windrows. This data can then be analyzed in the lab and used to adjust the composting process as needed and ensure the production of high-quality compost.

Quality control: Establishment of a lab is crucial in linking the decentralized facilities to a centralized facility. The lab controls the quality of the final product. Quality control involves the continuous evaluation and assessment of the composting process and the final product in order to ensure that it meets the required standards and specifications. Analysis of chemical properties defines stability (C:N < 20), the potential toxicity of the compost (e.g. heavy metals, organic pollutants), agronomical value (nutrients, pH, EC) and maturity (toxicity to plant root). A stable and mature final product should meet the criteria set by national standards.

Branding and Labelling: The final product is sold either in bulk or in bags. Preferably the final product should have a moisture content of lower than 40% before packing, to avoid water transportation with the compost. Packing the materials should be done before it is out on the market. The proper packing choice is waterproof, porous bags made of woven polypropylene. The concept suggests the development of a unique brand name and logo for the final product. The bags should contain information about raw feedstock, weight, estimated nutrient content, data on use and the ratio of application for different plants.

Certification: The certification process is typically carried out by independent laboratories that specialize in the testing and analysis of compost. However ongoing monitoring and testing onsite make it easier to meet the needs of customers and regulatory agencies.

Marketing: When compost production is part of the waste management system and subsidized by the local government the marketing strategy is mainly based on the demand and the potential application. In this respect the identification of a potential customer, willingness to pay, existing markets, competitors and their prices are crucial. The quality of the final product and possible application should be clearly defined. Labelling and certification of the product is a crucial strategy to reach better marketability. The existing markets for fertilizers and potential competitors should be examined carefully. This would ease the market evaluation for potential customers and the pricing of the product. In the beginning only cost recovery of the process may be viable. Deciding the distribution channel is another decisive factor based on the market examination. Direct delivery or selling through a local retailer with a good network of customers are possible choices. Lastly the promotion of the product through marketing strategies enhances customer trust and some additional examples are advertising, free delivery and application instruction manuals.

Training center: There is a direct correlation between a lack of skilled manpower and a system malfunction. In many cases the responsibility is assigned to the waste management staff and personnel without proper training. A training center is considered for the capacity-building of workers at decentralized and centralized facilities. The program enables the workers to assess waste received at the plant including weighting and characterization, handling of fine and coarse fractions after processing, O&M of plant machinery, using personal protective equipment, Health, Safety and Environment cautiously.

7.4.3 *Financial Overview for Designing the Composting Facility*

Details of the construction and operation of a centralised windrow composting plant with 10,000 tons of receiving waste per year are shown in the above table. In the construction year a total amount of 1.7 million euros should be invested in the civil engineering and purchase of the materials and equipment. Considering 310 working days in a year at least 1 manager and 6 workers are required to operate 32 tons of feedstock per day. The facility operation cost adds up to about 187,000.00 euros per year which also includes the capital related costs with an interest of 6% p.a on the additional investment as loan. The construction of a windrow composting plant can be invested through a government grant, loan or equity investment. Selling compost and receiving gate fees are the main sources of generating revenue. Considering the sale of compost at 58 euros per tons there could be around 186,669.00 euro can be generated as revenue every year. Considering 30% of total investment subsidising from the government and the rest 70% of the capital cost is obtained as a loan, there is a possibility to achieve amortization under 7 years. However the profitability of the entire composting plant mostly depends on the marketability of the compost produced. It is also important to note that the cost information will vary depending on whether the site in consideration is an existing composting plant or a proposed site, as well as the degree to which the planning, resource consent processes and design have progressed. The prices shown are for purposes of illustration only and are subject to change based on a variety of factors, including those related to the local market and economy as well as those particular to the site-specific data. See Table 7.8 for the financial detail of the designing and operating of a centralised compost facility.

7.5 Centralised Facility to Manage Recyclable Materials (MRF)

Recycling and recovery of waste involves converting waste materials into substances, products or materials through a reprocessing process. Although recycling rates in Senegal have increased there is still a lack of comprehensive data comparing the country's waste management performance to other African nations. There are only a few wastes reduction, re-use and recycling initiatives operating at scale across the entire country. Most waste recovery and recycling facilities are owned by private entities and are primarily concentrated in Dakar and its surrounding region. Recyclables waste from other municipalities must be transported to urban centres where processing and marketing capacity exist to meet export standards. It is imperative that the waste community places utmost importance on enhancing the quality of recycling results alongside boosting recycling rates. This must be the primary objective of all future endeavours to achieve optimal environmental benefits of recycling practices in Senegal.

Table 7.8 Financial overview for designing and operating of the Composting facility

Exchange rate			
Local currency		CFA	
1 Euro =		655.88	Ref. 2003 (www.xe.com)
Cost estimation of composting plant (size: 1000 t/a)			
Component	Investment %	Cost in Euro	CFA
Estimated civil engineering investment cost (part 1)			
Preparation of land and layout	0%	6000	3,935,280
Building construction	29%	500,000	327,940,000
Outdoor facility	41%	700,000	459,116,000
Other	6%	100,000	65,588,000
Sum of part 1	76.87%	1,306,000	856,579,280
Investment cost on material and equipment (part 2)			
Backhoe loader (JCB 1 CX)	6.95%	118,000	77,393,840
Rotary drum screen (Berkili Siebmaster 2)	3.53%	60,000	39,352,800
Grinder (Willibald UZ80)	3.53%	60,000	39,352,800
Bag opener	2.10%	36,000	23,611,680
Agethen conveyor belt 12 m	1.41%	24,000	15,741,120
Trommel sieve and magnetic separator	4.71%	80,000	52,470,400
Small tools (temperature measuring lance, chainsaw etc.)	0.29%	5000	3,279,400
Office equipment	0.59%	10,000	6,558,800
Sum of part 2	**23.13%**	**393,000**	**257,760,840**
Total setup costs	**100%**	**1,699,000**	**1,114,340,120**
Investment cost management			
State grants	30%	509,700	334,302,036
Own contribution	0%	0	0
Additional investment required	70%	1,189,300	780,038,084
Yearly costs for operation and maintenance			
Personnel cost		20,000	13,117,600
Energy consumption		22,011	14,436,574.68
Water		–	–
Equipment and others		20,000	13,117,600
Site management		13,300	8,723,204
Building maintenance		5000	3,279,400
Machine maintenance		30,000	19,676,400
Insurance		5000	3,279,400
Total operation and maintenance cost/year		**115,311**	**75,630,178.68**

(continued)

Table 7.8 (continued)

Exchange rate			
Local currency		CFA	
Yearly cost for capital, consumption and operation			
Capital related cost		71,358	46,802,285.04
Operation and maintenance costs		115,311	75,630,178.68
Total yearly costs		186,669	122,432,463.72
Yearly profit from the sales of compost at 5000 €/ton			
Compost sales		380,683.04	32,358,058.40
Total profit/year		380,683.04	32,358,058.40
Net profit		265,372.04	22,556,623.40
Amortization year	*< 7 years*		

A region wise centralized recycling plant in Senegal can have a significant positive impact on the environment and economy by managing energy recovery waste which is now not properly disposed. Long-distance domestic waste movement is made more feasible by converting combustible waste at the material recovery facility (MRF).[14,15]

Economic and Environmental Benefits:

- **Waste Reduction**: The facility would enable the reduction of a substantial portion (20%) of total generated waste as leachate production in the event that the waste is placed in a landfill.
- **Greenhouse Gas Emissions Reduction**: Waste recycle helps prevent the release of methane a potent greenhouse gas. Methane emissions contribute to climate change.
- **Resource efficiency**: The pre-sorting including household and decentralised composting would reduce the waste amounts to be treated and transported to the centralized facility. This would indirectly result in the overall efficiency by saving fuel and other resources. The decentralized facilities should have a storage of certain volumes of segregated RDF materials in the form of bales regulating and easing periodical collection within the individual zones/regions.
- **Job Creation**: The establishment and operation of a centralized recycling facility would create employment opportunities in various sectors, such as waste management, bale production, facility maintenance and transportation create jobs at the decentralized facilities and also at the centralized treatment and processing units/plants.
- **Revenue Generation**: The facility can generate revenue through the sale of baled waste or in the form of pellets. The baled/pelletized RDF can be easily shipped to more locations by being compressed into bales/pellets. This revenue

[14] Swachh Bharat Mission Municipal Solid Waste Management Manual (2016).

[15] https://www.epa.gov/facts-and-figures-about-materials-waste-and-recycling/municipal-solid-waste-generation-recycling-and. Reterived on 20.05.2023.

stream can contribute to the financial sustainability of the facility and whole waste management operations.
- **Resource Recovery**: Converting combustible waste into RDF increases the product's calorific value and can be sold as fuel for cement industries or other industries. Further depending on the quantities produced, a waste to energy incineration plant can be setup.
- **Waste Management Cost Reduction**: Proper management of combustibles can potentially reduce the costs associated with transportation and landfill disposal. This cost reduction can benefit both municipal authorities and waste generators as well as the waste management and processing companies.

A conventional Material Recovery Facility (MRF) is often situated within a structure resembling a warehouse, characterized by the presence of concrete flooring and contained by a perimeter fence to ensure security measures. The facility should consist of the following elements:

- an area designated for receiving or tipping,
- a section dedicated to sorting and processing,
- a storage space for recyclable materials,
- a designated area for storing residual waste,
- a zone for housing equipment,
- a designated office space, and
- a loading area specifically designed for handling both residual waste and processed recyclable materials.

It is necessary that the facility is equipped with essential infrastructure for water and electrical connections as well as sufficient room to accommodate the entry and exit of trash vehicles. The inclusion of facilities for washing and a septic tank is vital. The design of the warehouse will aim to minimize the positioning of columns that may impede the smooth flow of products and equipment, while also enabling the installation of taller ceilings. According to a study by the Asian Development Bank[16] it is recommended that receiving areas possess the capability to accommodate a minimum of two days' worth of the processing capacity of the Material Recovery Facility (MRF). This is to account for any equipment malfunctions and to ensure the availability of materials for the second-shift operation.

MRFs that are operated manually and have a capacity of less than 2 tons per day often possess roofed floor areas measuring at least 50 m^2. These areas are solely dedicated to housing the receiving, processing and storage sections of the facility. Facilities transitioning from semi-automated to completely mechanized operations will necessitate areas ranging from 150 to 1500 m^2 with the exclusion of parking and buffer zones. According to the Material Recovery Facility Handbook published by the Recycling Marketing Cooperative of Tennessee in 2003, it is recommended that MRFs processing less than 10 tons of recyclable waste per day have a building area

[16] Materials recovery facility tool kit, Asian Development bank.

of no more than 1400 m^2. For facilities that handle waste not exceeding 100 tons per day, it is suggested to have a floor area of approximately 1800 m^2.

7.5.1 Detailed Process Description of a MRF Facility

The fundamental apparatus for a MRF manual process would comprise of sorting tables, weighing scales, a baler and a payloader. Semi-automated MRFs employ a conveyor system that can be enhanced by the inclusion of a loader to streamline the sorting process. Automated facilities employ a range of equipment, including screens, magnetic separators, air classifiers and conveyor systems which can be configured to accommodate several processing lines. The selection of equipment is contingent upon the desired capacity as well as the characteristics and composition of the incoming waste stream.

The process of recycling municipal solid waste (MSW) involves several steps that aim to effectively sort, process, and recycle various materials from the waste stream. Here's a detailed description of the process at MRF facility can be found in Table 7.9 with the described in the 8 steps.

- **Waste loading to conveyer**: MSW is collected from households, businesses and public areas and transported to a recycling facility. Upon arrival the waste is unloaded and directed to a sorting area where it undergoes an initial separation process. This can involve manual sorting by workers or automated equipment such as conveyor belts and sorting screens. The goal is to remove large items, bulky waste and potentially hazardous materials.
- **Bulky material separation through trommel screen**: Once the preliminary manual sorting is complete, the waste is often sent through trommel screen to separate bulky material and increase processing efficiency. Trommel screen separates larger items like furniture appliances and tree branches into smaller, more manageable pieces. This step also helps in liberating valuable materials for further processing.
- **Metal Separation**: after removing the bulky materials, the waste stream passes through magnetic separators. These devices use powerful magnets to attract and separate ferrous metals such as steel cans and iron-containing objects from the waste stream. An eddy current separator repels the aluminium cans and foils from the conveyor. The separated metals are collected for recycling while the remaining waste continues along the recycling process.
- **Screening and Air Classification**: The shredded waste may undergo further screening to separate it based on size. Trommel screens which are cylindrical screens with different-sized holes are often used. The screens allow smaller particles to fall through, separating them from larger materials. Air classifiers may also be employed to separate lightweight materials like plastic film from heavier recyclables.

Table 7.9 Different process or stages and equipment employed in a material recovery facility[a]

Process or stages	Equipment
Pre-sorting material handling equipment	Belt conveyor Screw conveyor Apron conveyor Bucket elevator Drag conveyor Pneumatic conveyor Vibrating conveyor Debagger
Ferrous metal separation	Magnetic separator and screening
Screening	Disc Screening Trommels
Air classification	Horizontal air classifier Vibrating inclined air classifier Inclined air classifier
Non-ferrous metal separation	Rotating disk separator Eddy current separator
Size reduction	Can densifier Can flattener Glass crusher Plastic granulator Plastic perforator Baler
Pollution control	Dust collection system Noise suppression devices Odour control system Heating, ventilating, and air conditioning (hvac)
Other fixed equipment	Fixed storage bin Live-bottom storage bin Floor scale for pallet or bin loads Truck scale Belt scale

[a] Swachh Bharat Mission Municipal Solid Waste Management Manual (2016)

- **Optical Sorting**: In this step advanced optical sorting systems are used to identify and separate different types of materials based on their optical properties. Optical sensors and cameras detect specific characteristics such as colour, shape and reflectivity to sort materials like plastics, paper, cardboard and other recyclables. This technology helps increase the efficiency and accuracy of the sorting process.
- **Compaction and Baling**: Recyclable materials such as cardboard, paper and plastics, are often compressed into compact bales using balers. This reduces the volume of the materials, making them easier to store, transport, and handle. Compactors may also be used to reduce the volume of non-recyclable waste that will be sent to landfill.

- **RDF production/energy recovery**: The remaining waste consisting of non-recyclable and residual materials with higher calorific value undergoes further processing to produce RDF. The waste is typically subjected to additional shredding, screening and refining steps to remove contaminants, improve homogeneity and increase the calorific value of the resulting RDF. In some cases, the RDF may undergo a drying process to reduce its moisture content. This is done to enhance the energy efficiency and combustion properties of the RDF.
- **Transport and Distribution**: The sorted and processed recyclable materials and RDF are stored in designated areas before being transported to recycling facilities or buyers. Proper storage and organization help streamline the recycling process and ensure efficient distribution to industries that will utilize these materials as feedstock for new products.
- **Rejects sent to landfill**: Final rejects obtained after the material and resource recovery from the waste can be collected and sent the landfill site.

7.5.2 *Overview of the Size of the Region and Zone/Region-Wise Centralized Recycling and Material Recovery Facility in Senegal*

To provide an overview of the size of the district and region wise MRF facility, that has to be considered during the planning phase, Table 7.10 highlights the data and Fig. 7.7 showcases the potential locations and number of required facilities. The estimation has been made considering the amount of recyclable waste that is generated in Senegal with the projected population till 2050. The approximate size of the district and region-wise MRF facility is estimated by considering 1800 m^2 of area to treat 100 tons of recyclable waste reaching the facility. These areas are solely dedicated to housing the receiving, processing and storage sections of the facility with the exclusion of parking and buffer zones. Based on the analysis the map (Fig. 7.7) displays the most appropriate areas for placing waste MRF facilities estimated to treat the generated amount of MSW in each region. The capacity of the MRF facility is calculated at 30,000 tons of recyclable waste generated annually. It was found that the Dakar region would require more than 10 MRF facilities to treat the estimated 6 million tons of recyclable waste generated annually in 2050, while the region of Kédougou would only require one MRF. About half of the 14 regions would require between 3–4 MRF facilities with a capacity of 30,000 TPA.

7.5.3 *Financial Overview for Designing the MRF Facility*

Details of the construction and operation of a centralised MRF with capacity of 20,000–26,000 tons per day of recyclable MSW are shown in the Table 7.11. It's important to highlight the labor cost which makes up 42% of the annual operational

Table 7.10 Overview of the size of the centralized-material recovery and recycling bailing facility in Senegal

Zone	Region	Population in 2050 by region	Approximate area required to set up centralised Recycling plant (ha)	Approximate area required to set up centralised Recycling plant in (in Million Euro)
Dakar	Dakar	8,978,949.95	829.80	6.48
Cap Vert-Thies	Thies	5,065,556.28	351.10	2.74
Central Senegal	Kaolack	2,814,404.27	195.07	1.52
	Diourbel	4,399,975.91	304.97	2.38
	Fatick	2,145,569.71	148.71	1.16
	Kaffrine	1,737,654.41	120.44	0.94
Total area in acres			769.20	6.01
Northern Senegal	St. Louis	2,554,620.08	177.07	1.38
	Louga	2,491,439.95	172.69	1.35
	Matam	1,753,083.49	121.51	0.95
Total area in acres			471.26	3.68
Tambacounda	Tambacounda	2,081,752.07	144.29	1.13
	Kédougou	452,527.66	31.37	0.25
Total area in acres			175.66	1.37
Casamance	Kolda	1,945,093.93	112.38	0.88
	Ziguinchor	1,621,432.13	94.32	0.74
	Sédhiou	1,360,870.49	112.38	0.88
Total area in acres			341.53	1.67

expenses. Two significant factors determine the personnel cost; the facility's capacity and the minimum working hour tariff which varies across different countries. Detailed labor costs can be seen in Table 7.11 to imagine how this component taking the vast majority of the annual operation cots of MRF.

The economic feasibility of adding an MRF depends on its capital and operating costs. Assuming a 20-year design life, the facility must process the required tonnage. There are two main components that make up the capital costs. The first one is the construction cost which includes design and engineering, soil reports and permitting, construction management, construction costs and general contractor fees. The second component is the equipment and processing system as seen in Table 7.12. To account for any unknown issues that could arise during the project a planning-level contingency of 25% was included. For capital cost estimating purposes some of the key features below are assumed.

- Fully enclosed, pre-engineered metal building
- High-strength concrete floor with wear-resistant features

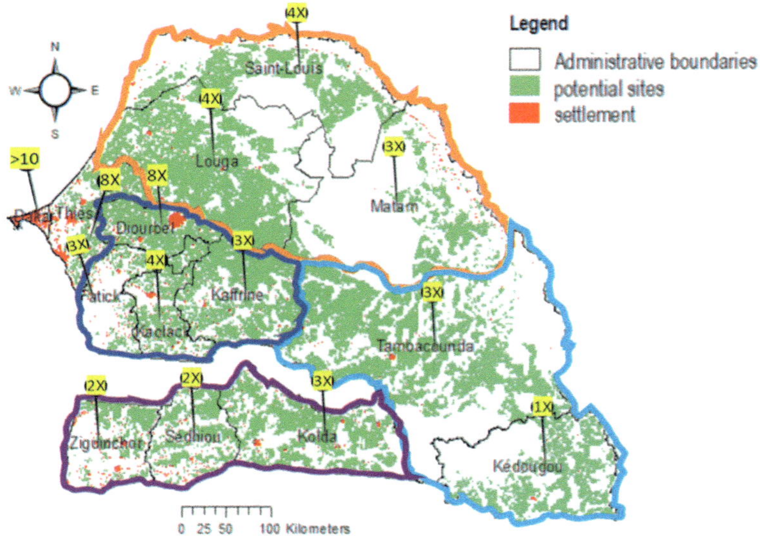

Fig. 7.7 Potential MRF sites and number of facilities required in each region of Senegal

Table 7.11 Annual labour costs for MRF

Category	Labor cost rates (EUR/hour)	Full-time employee equivalents	Annual cost (EUR)	Annual cost (1 EUR to 655.9 CAF)[a]
Facility manager	74	0.5	78,000	51,158,640
Marketing	41	1	84,000	55,093,920
Admin/accounting	41	0.5	42,000	27,546,960
Mobile equipment operators	30	2	123,000	80,673,240
Shift foreman	47	1	100,000	65,588,000
Baler/recycling equip. operators	30	1	60,000	39,352,800
Materials sorters	20	10	420,000	275,469,600
Maintenance/ mechanics	34	2	140,000	91,823,200

[a] Ref: 2023 (https://www.xe.com/currencyconverter/convert/?Amount=1&From=EUR&To=XOF)

- Building electrical with a combination of natural lighting and conventional high-intensity
- Overhead lights
- Mechanical exhaust system
- Drive through the unloading area
- High bay, column-free receiving area

Table 7.12 Financial overview for designing the MRF facility

Exchange rate			
Local currency		CFA	
1 Euro =		655.88	Ref. 2003 (www.xe.com)
Cost estimation of MRF plant (size: 20,000–26,000 t/a)			
Component	Investment %	Cost in Euro	CFA
Estimated civil engineering investment cost (part 1)			
Design and engineering, soil reports and permits, construction management	9%	2,500,000	
Construction costs, including construction and general contractor fees	44%	12,000,000	
Other	1%	140,000	
Sum of part 1	54.04%	14,640,000	
Investment cost on material and equipment (part 2)			
Process system	2.21%	600,000	393,528,000
Equipment installation, start-up and contingency (30%)	6.64%	1,800,000	1,180,584,000
Mobile equipment	1.85%	500,000	327,940,000
Mobile equipment contingency (10%)	0.18%	50,000	32,794,000
Total equipment	35.07%	9,500,000	6,230,860,000
Sum of part 2	**45.96%**	**12,450,000**	**8,165,706,000**
Total setup costs	**100%**	**27,090,000**	**17,767,789,200**
Investment cost management			
State grants	30%	8,127,000	5,330,336,760
Own contribution	0%	0	0
Additional investment required	70%	18,963,000	12,437,452,440
Yearly costs for operation and maintenance			
Personnel cost	42%	1.050.000	688.674.000
Energy consumption	23%	585.000	383.689.800
Water		–	–
Equipment and others	14%	350.000	229.558.000
Residuals haul and disposal	6%	160.000	104.940.800
Contingency	10%	250.000	163.970.000
Insurance	5%	130.000	85.264.4000
Total operation and maintenance cost/year		**2.525.000**	**1.656.097.000**

(continued)

Table 7.12 (continued)

Yearly cost for capital, consumption and operation			
Capital related cost		1.137.780	746.247.146
Operation and maintenance costs		2.525.000	1.656.097.000
Total yearly costs		**3.662.780**	**2.402.344.146**
Yearly expected profit from the sales of materials			
	Net market revenue (EUR/ton)	Total revenue	Tons separated
ONP (old newspaper)	24	99.131	4.130
Mixed paper	29	317.500	10.948
OCC (old corrugated containers)	57	264.000	4.632
Cartons	38	3.000	79
Steel/tin cans	280	200.000	714
Plastics #1 PET	270	307.000	1.137
Plastics #2 HDPE natural	750	375.000	500
Plastics #2 HDPE color	290	130.000	448
Plastics #3-#7	10	700	70
Aluminum UBCs	15,000	375.000	250
Glass	5	15.000	3.000
Total profit/year		2.086.311	25.909
Net profit		– 438.669	
Net operational costs			
Amortization year	*< 10 years (assumed)*		

- Concrete containment wall in the receiving area
- Bale storage area
- Four loading docks
- Concrete roadways and approaches
- One inbound weigh scale on the approach road
- Area for office, conference room and locker room/restroom facilities
- Elevated viewing gallery
- 10% general contractor fees
- 25% contingency
- 20% development costs of design/engineering, soil investigations, permitting, and construction period services.

MRF processing equipment included in the financial estimation are: One processing system line which is anticipated to include infeed conveyor, flow metering, presort station, cardboard screen, glass screens, fiber/container separation screen, fiber optical sorting, ferrous magnet, container optical sorting for selected commodities, ECS, QC stations and live bottom bunker and bin storage There are two balers for recyclable material. Calculation considered 20% for process equipment installation and start-up and 10% contingency. See Table 7.12 for the financial detail of the designing and operating of a centralised MRF.

7.5.4 Approach and Timelines for Implementing Centralized Recycling Treatment Facilities

To improve the MSW treatment situation in Senegal through centralized facilities the first step would be to control the incoming materials. To achieve this there should be various interlinked promoting and awareness rising activities such as promoting animal feeding network, encouraging home composting, and setting up decentralized waste facilities with storage possibilities. This will help decrease the burden on the formal MSW management system. To ensure that the transported waste is received in a timely and efficient manner, the payloads of the vehicles using a weighbridge or visual estimation have to be carried out at the decentralized and centralized facilities. It's important to process the receiving waste daily to minimize local disruption and unintended accumulation choking the facility and also causing environmental impacts. Facility cleaning, leachate management, baling, pelleting, sand cleaning, etc. of the recovered materials should also be done regularly. Integrating the informal sector into the pre-sorting process would also be beneficial and enhance the whole system further. Lastly we should focus on capacity building and training for all manpower in the facility to ensure optimal operation. The phase of the strategy to be implemented are suggestive and described in Fig. 7.8 taking into consideration the effect of on waste composition.

First Phase, Existing situation analysis and optimization, and identification of location: In order to achieve short term goals within the next two years it is vital to prioritize the analysis and optimization of existing decentralized waste treatment units as well as identifying locations for centralized treatment facilities. To accomplish this a digitalized map should be created to show active and inactive waste treatment and disposal in all of Senegal. This map must include information on land availability centralized or decentralized processing units, landfill areas or dumpsites and more. Each processing unit should be described in detail including waste processing techniques, percentage of waste treated, capacity for gas and electricity generation and nearby end disposal facilities. Further an environmental impact assessment should be conducted on a state level to evaluate new facility locations. This assessment should take into account projected waste amounts, future urban development and the protection of natural resources such as water bodies. Finally controlled processing facilities should be established and market potential for treatment facility products should be surveyed.

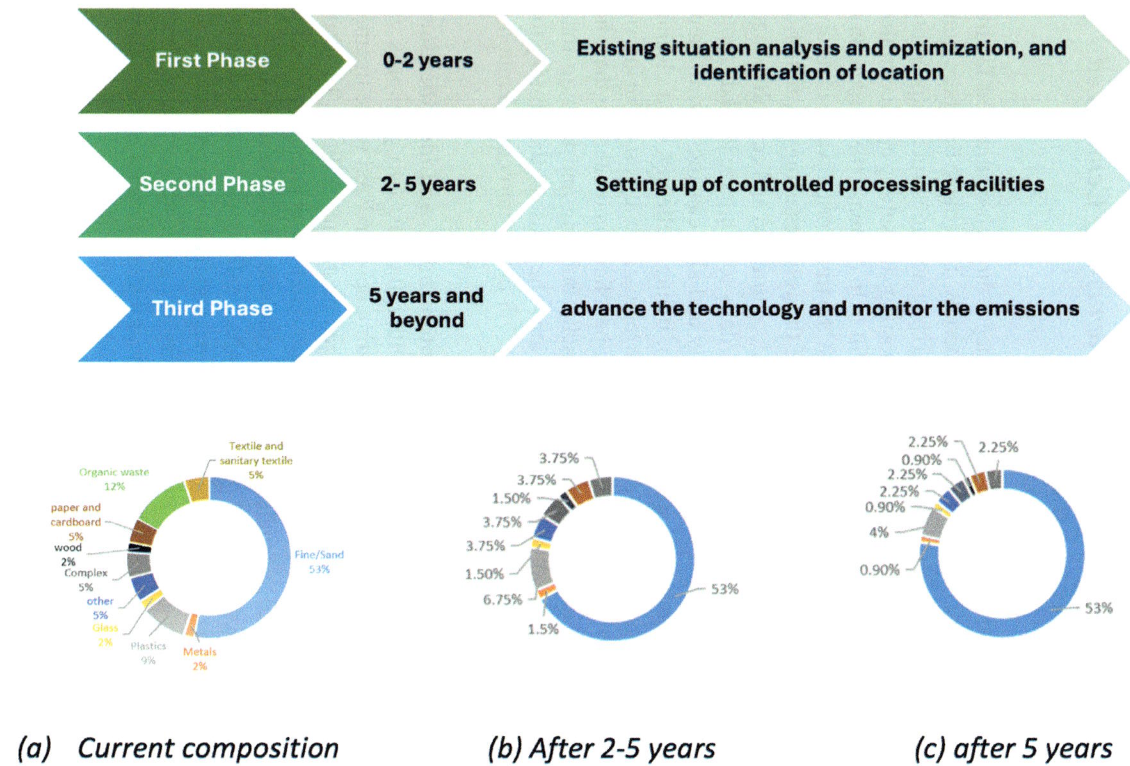

Fig. 7.8 **a** Step by step enhancement of waste treatment and applying the centralized strategy; **b** shows the composition change after involving the MRF; **c** waste composition after involving incineration with 40% loading rate

Second Phase: Setting up ccentralized treatment facility: In the next few years it is important to focus on designing and selecting a treatment facility. A committee chaired by the sanitary inspector should consider several factors such as the daily processing capacity, the quality of materials used for building the facility, the efficiency of the chosen technology and its certifications and the availability of nearby storage facilities. Additionally composting and MRF units should be constructed taking into account the technical aspects of the system and future waste generation for at least 40 years.

Third Phase: Operating, maintenance, advancing the technology and monitor the emissions: When it comes to ensuring stable operation, financial sustainability and longevity of treatment facilities have to be kept in place and maintenance should be delegated by trained workers. It is also important to consider key factors in the establishment of MRF facilities such as land ownership, waste collection and transportation routes, location, waste sorting, processing and storage facilities. In addition the designing, developing and constructing advanced treatment facilities must be considered to add value to the sorted products such as RDF or Biogas.

The proposed strategy as centralized treatment streams have been illustrated previously in Fig. 7.5 and the effect on the waste composition can be explained further here. Around 25% of the daily generated waste can be separated into its essential fractions in MRF plants it was assumed that 12.5% of the materials can recycled and 12.5% for RDF production. The organic fraction (12%) would be managed and treated by the decentralized treatment streamlines (as explained in Chap. 5) and centralized treatment facilities like biogas plants and composting projects. The effect of this scenario takes out 25% from each fraction of the waste, meaning that for the case of metals which have 2% as a share in the waste composition will be reduced by 25–1.5% and so on with other fractions, where plastic, glass, other, complex, wood, paper and cardboard, and textile will be reduced to 6.75, 1.5, 3.75, 3.75, 1.5, 3.75, 3.75% respectively. This milestone can be achieved during the second phase of the strategy plan by involving the MRF facility and starting the operation mode. Coming back to the total fraction again, now 25% can be separated in MRF to produce RDF and recycle the other valuable materials, and the 12% organic fraction can be treated in a parallel manner by applying the concept of centralized and decentralized treatment streamlines. The rest 63% of mixed waste can be divided into two fractions; around 20% will be converted to landfills, and at the same time around 40% will be treated in waste incineration causing more variations in the waste composition, as can be seen in Fig. 7.13. Accordingly, 40% will be reduced simultaneously in each waste fraction.

7.6 Centralised Waste to Energy Facilities (Energy Recovery)

7.6.1 Detailed Process Description of a WtE Facility

Waste-to-energy technologies (WtE) are promising technologies especially for developing countries to turn waste into a useable form of energy. WtE play a vital role for sustainable waste management and mitigation of environmental issues. This concept can be achieved through many different technologies by two main streams; either chemical-biological treatment of the waste means biogas plant or thermal conversion of waste by mass burning, gasification and pyrolysis.

One of the widespread technologies in Germany is the biogas plant but it must be always kept in mind that the biogas technology is designated for the organic fraction in other words the concept of separation at source should already be applied to ensure that the fuel required for the biogas plant (organic waste) can be continuously provided. In the case of countries that are recognized with mixed waste generation and waste heterogeneity a pre-treatment facility like MBT (Mechanical Biological Treatment) is required to secure the materials flow for the biogas plant. Energy utilisation of waste can be divided into several groups according to whether the waste is just incinerated or whether a fuel for later utilisation is produced. The production of energy from waste can take place either by thermal decomposition (gasification, pyrolysis, and hydrolysis) or by biological processes (anaerobic digestion, fermentation). The objective of thermal treatment of the waste is to lower its volume and to utilise the chemically bounded energy in the waste. The utilisation can be achieved by production of energy in the form of heat and/or electric power and by production of alternative fuels. In general thermochemical conversion processes have come more into focus during the recent decades because of their benefits such as higher conversion efficiency, zero-waste concept, lower resistance time, better economic performance and hydrogen of the biomass are oxidized to produce CO_2 and water respectively.

Gasification and pyrolysis are very advanced levels of thermal conversion of waste that are still in the research and development stage in well-developed countries like Germany and Japan; however, both technologies deal with a certain fraction of waste. For example, gasification can achieve good conversion efficiency if the biomass was the input fuel in the reactor. In addition, this technology consumes enormous amounts of water to achieve the gasification process. Thermal conversion of waste is more advantageous for countries that have mixed waste generation aspects and rely on the rich experience of the waste and resource management department at the University of Rostock and academic research (1:4 biomass to water ratio). Pyrolysis technology is more complicated than gasification and it's more applicable for plastic pellets or RDF (Refuse Derived Fuel). Technically is non- oxy or partial–oxy thermal conversion process of waste to energy meaning that this technology requires an advanced level of controlling process to achieve the conversion technique at the same time the manpower should also be well trained to operate such a facility. Apart

Table 7.13 Mass loading rate for the available waste-to-energy technologies

Technology	Typical application range (tones/day)
Moving grate (mass burn)	120–3000
Fluidized bed	36–200
Rotary kiln	10–350
Modular (starved air)	1–75
Pyrolysis	10–100
Gasification	250–500

from that both of them (gasification and pyrolysis) are very limited in terms of mass capacity due to their complexity and their selectivity for the waste fraction tends to be treated or converted into useful energy as can be seen in Table 7.13. Furthermore their reactors are designed to work at a high rate of temperature which means the process itself exploits massive amounts of energy. In comparison mass burning (direct waste incineration) has a long history in terms of the experience of waste conversion to energy for more than 200 years at the same time this technology has no complexity (classic combustion process) and no selectivity for waste fractions, in other words mixed waste is directly burned which is an optimum option for the developing countries who have mixed waste generation manner like in China. The mass capacity is unlimited meaning that for megacities (densely populated cities) more than 1 million inhabitants will be the best scenario. Note that well-developed countries like Germany still depending on the mass burning facilities in their systems. Right now more than 70 incinerators in Germany exist to accommodate different types of waste with huge quantities.

Pyrolysis

Pyrolysis is the thermal degradation of solid waste in the absence of oxygen. The burning of wastes is done in a controlled environment which is chemically known as complete combustion that releases water molecules and carbon dioxide into the atmosphere. External heat is required to maintain the temperature between 300 and 800 °C depending on the materials used in the process in this technology waste needs to be pre-treated. It requires the mechanical separation of glass, metals and inert materials. This process generates products such as syngas, biooil and char. The quality and yield of products depend upon various parameters such as pyrolysis temperature, heating rate, residence time, waste feedstock composition and particles size. The exhaust gases (syngas) produced in this process can be used to sustain the energy in the pyrolysis process. Thermal energy generated is used to produce electrical energy using a turbine and heat exchanger. It is important to clarify that pyrolysis can process biomass and plastic materials. Plasma pyrolysis is the highest advanced technology developed to produce syngas by transforming plastic waste with a high calorific value.[17]

[17] Moyaa et al. (2017).

Some advantages of the pyrolysis process are: the equipment is flexible for installation; there are minimum environmental issues; all waste materials are used to produce different bio-products; and the produced syngas can be used in different energy applications such as engines, boilers, fuel cells, turbines and heat pumps (see Table 7.14). The technology has many benefits as minimum oxidation of the waste, no production of pollutants, high level of synthetic gas production, high level of calorific values for this gas, clean gas production which can be used in Genset. Seems to be the most ecologically clean, simple and modular modern technology of waste utilization.

Gasification

Another thermochemical WtE conversion technology for the management of MSW is gasification. The Gasification process to produce energy from waste has been developed over the last three decades. In this process the MSW feedstock is heated at 500–900 °C in an adequately monitored oxygen/steam/air atmosphere. In this process part of the energy is used to produce heat to gasify the remaining fuel. This process involves partial oxidation and its main product is fuel gas (syngas). The constituents of the produced syngas are CO, H_2, CH_4, and trace quantities of ethane, ethane and ethylene. Gasification can reduce the waste mass by 70% and over 90%

Table 7.14 Advantages and disadvantages of WtE technologies[a,b,c]

Technology	Advantages	Disadvantages
Incineration	– Most suitable for high calorific value waste – Units with high throughput and continuous feed can be set up – Thermal energy for power generation or direct heating – Relatively noise less and odourless – Low lands are required – Can be located within city limits, reducing transportation costs – Hygienic	– Least suited for aqueous, high moisture content, low calorific – Value and chlorinated waste – Toxic metal concentration in ash, particulate emissions, SOx, NOx, Chlorinated compounds, ranging from HCL to dioxins – High capital and O&M costs – Skilled personnel required – Overall efficiency for small power stations is low
Pyrolysis/ gasification	– Production of fuel gas/oil, which can be used for various applications – Control of pollution superior as compared to incineration	– Net energy recovery may suffer in waste with excessive moisture – High viscosity of pyrolysis oil may be problematic for its burning and transportation

[a] Moyaa et al. (2017)
[b] Münster and Lund (2010)
[c] Sustainable management of municipal solid waste through waste-to-energy technologies. https://doi.org/10.1016/j.biortech.2022.127247

of the volume of waste. The syngas produced from gasification process is energy-rich and clean enough to produce energy in a gas turbine or engine and can be used for speciality chemicals, and liquid fuel manufacturing via various routes. As this process generates hydrogen as a clean energy source with a heating value of about 141.7 MJ/kg, it is considered a promising WtE technology.[18]

Incineration

The incineration of waste is one of the most widespread ways of thermal processing of waste that can be applied to a wide circle of different types of waste. Incineration is typically the combustion of waste at high temperatures (800–1200 °C) in the presence of excess oxygen or air. Incineration is a prominent waste treatment process that reduces waste bulk volume by 70–90% and is appropriate for wastes with higher calorific values. The combustion of MSW occurs via sequential steps including heating/drying, devolatilization/decomposition and combustion of volatiles and char. During the thermal degradation process to generate major gaseous components including water vapour, carbon dioxide, carbon monoxide, hydrogen and methane. The produced volatiles and char oxidise in the to produce heat. In this process the carbon and hydrogen of the wastes are oxidized to produce and water respectively. While incineration helps recover the energy from waste it also leads to the generation of greenhouse gases. However some efforts to control these concerns have resulting in regulations to reduce the amount of specific emissions such as: sulphur dioxide, hydrogen fluoride, nitrogen oxide and nitrogen dioxide, total organic carbon, hydrogen chloride, carbon monoxide, dust and volatile heavy metal.[19] On the other hand some advanced technologies have been developed to reduce these effects.

Table 7.14 shows the advantages and disadvantages of the WtE technologies showing their similarities and differences.

7.6.2 *Importance of Waste Composition and Lower Calorific Value for Waste to Energy Incineration*[20]

Waste-to-energy (WtE) incineration is distinct from power generation through the use of fossil fuels in that the amount of power generated is contingent on the characteristics of waste, particularly lower calorific value (LCV). It is possible that the facility will not be able to generate the anticipated quantity of electricity or that the facility itself will be rendered inoperable if the waste does not satisfy certain properties. Additionally, the composition of the waste as well as its LCV shifts over time

[18] Münster and Lund (2010).

[19] Sustainable management of municipal solid waste through waste-to-energy technologies. https://doi.org/10.1016/j.biortech.2022.127247.

[20] CCET guideline series on intermediate municipal solid waste treatment technologies (based on Kawai (2016) and World Bank (2018).

as a result of seasonal shifts in the amount of waste that is produced. For instance, during the rainy season, waste might have a higher concentration of moisture. These are the kinds of considerations that need to be taken into account when designing the furnace and the overall process of loading waste. The LCV must be at least 7000 kJ/kg on average and must never fall below 6000 kJ/kg for waste energy incineration to be successful in terms of energy recovery. The LCV is closely related to the composition of the waste they produce. The use of a triangle diagram was suggested by Kawai and Tasaki (2016)[21] in order to determine whether or not the current proximate composition of municipal solid waste, which includes moisture, ash, and volatiles, would be suitable for incineration, composting, and the production of RDF. It is demonstrated by the triangle diagram that the range of techniques that are applicable can be roughly identified by the composition of the waste itself. For instance, the technology of incineration can be utilized for waste that has a moisture content of 75% or less and a volatile content of 20% or more, which is equivalent to a low-capacity value (LCV) of 3352 kJ/kg or higher. Furthermore, in situations where energy recovery is a component of the incineration process, these technologies are typically applicable to waste that has a volatile content of 30% or more, a moisture content of 65% or less, and a latent heat capacity of 6285 kj/kg or more.

Waste composition is influenced by many factors, such as the level of economic development, cultural norms, geographical location, energy sources, and climate. In general, as a country urbanizes and populations become wealthier, the consumption of inorganic materials increases, while the relative organic fraction decreases. If the LCV (Lower Calorific Value) falls below 6000 kJ/kg, for example 4500 kJ/kg, or even lower with auxiliary fuel, incineration would be still possible, however, it would not be efficient for WtE incineration. In general, waste in developing countries contains a high percentage of food and kitchen waste and a LCV than that in developed countries. To reduce moisture content in combustible waste, developing countries should focus on source separation and reducing the moisture content of food and kitchen waste. This aspect is clearly mention in the strategy for handling the waste at source, strategy for proper waste collection and transport and strategy for centralized treatment for organic waste. The implementation of the developed strategies in Senegal, which focus on managing organic waste at its source, establishing a proper waste collection and transport system, and creating a centralized treatment facility for organic waste, will significantly impact the potential for incinerating waste with a higher LCV without the need for additional support. This approach also offers greater economic incentives.

[21] Kawai and Tasaki (2016).

7.6.3 Overview of the Size and Finances of a Centralized WtE in Senegal

As other facilities gasification consists of capital and operation costs. According to the literature and available facilities in reality the fixed modules cost and its auxiliary work occupy around 90% of the capital cost, while design and engineering, soil reports, permitting and construction management and costs are occupying around 10%. As an approximation the capital cost ranges between 3 and 5 million Euros depending on the end products of the facility and its designed capacity (see Table 7.16). The major components that have the highest share of the capital module costs are the gasifier and the liquid–gas separator with 17 and 35% respectively. Operation cost is around 500,000–700,000 Euros, it can be said that the operation cost can be taken as 14–15% of the capital costs. The pyrolysis technology has the same financial aspects (see Table 7.17).

Mass burning

Waste incineration is a major component of any well-developed waste management system and plays a main role in two sectors in waste sector as a treatment facility and in energy sector as a conversion plant. Waste incineration is well-recognized for its massive capacity (accommodation of waste), is technically approved, and its ability to receive heterogeneous materials due to that this technology is widespread in comparison with gasification and pyrolysis. Its high capacity increases its cost, in Germany, the capital cost of a waste incineration plant is around 500 Euro/ton as a rule of thumb the operational costs will be 10–15% of the capital costs. For the capital modules around 50% of the capital cost is related to the waste handling in the bunker, crane, crane grate incinerator grate (furnace) and part of the Rankine cycle like the boiler (economizer, evaporator and superheater). On the other hand, 17% is designated for the flue gas cleaning system starting with the reduction of nitrogen oxides NO_x by using selective catalytic reduction or non-selective catalytic reduction which needs ammonia NH_3 as an additive compound to reduce the nitrogen oxides in turn this needs to be calculated in the form of continuous and variable operational costs. The same issue in the case of sulphur dioxide reduction, HCl, HF and particles in the fly ash, where sulphur dioxide can be eliminated by using calcium hydroxide.

For Senegal, a standard waste incinerator capacity of 200,000 tons per year has been proposed. Based on this capacity and an average lower calorific value of 6.5 MJ/kg for mixed waste, approximately 89,000 MWh of energy can be recovered annually by each facility (see Table 7.15).

Dakar, the main city and region, is populated by over 4 million inhabitants who generate 1.18 tons of mixed waste per year. Most industrial activities are also concentrated in Dakar. However, since Dakar is a harbor city and serves as Senegal's gateway to the world, the construction of waste incineration plants or other facilities (such as materials recovery facilities) would be logistically challenging due to transportation constraints. Therefore, Thies, Touba, and Kaolack have been selected as locations for waste incineration plants to process daily waste generation and recover energy.

Table 7.15 Overview of the size and number of the centralized incineration facility in Senegal

Regions	Waste generation in million TPA	Energy consumption (273 kWh/cap.year)	Waste incineration potentiality (200,000 ton/year) for each plant	Expected power produced from each plant (kWh/year)	Expected power produced for the proposed distribution (kWh/year)	People's consumption power covered by incinerators (cap)
	2022	**2030**				
Dakar	1.18	1,103,527,425				
Thies	0.42	622,565,034	3	89,013,000	267,039,000	978,164.84
Kaolack	0.23	345,894,822	2		178,026,000	652,109.89
Diourbel	0.36	540,764,133	2		178,026,000	652,109.89
Fatick	0.18	263,693,976	–		–	–
Kaffrine	0.14	213,560,529	–		–	–
St. Louis	0.21	313,966,926	–		–	–
Louga	0.20	306,201,987	–		–	–
Matam	0.14	215,456,787	–		–	–
Tambacounda	0.17	255,850,686	–		–	–
Kédougou	0.04	55,616,379	–		–	–
Kolda	0.16	239,055,180	–		–	–
Ziguinchor	0.13	199,276,623	–		–	–
Sédhiou	0.11	167,253,177	–		–	–

The proposed distribution of incinerators is determined by each city's population and waste generation rates (see Fig. 7.9). Three incinerators have been proposed for Thies, each with a capacity of 200,000 tons per year. Power can be supplied to nearly 1 million inhabitants by these three facilities, with the energy needs of approximately 25% of Dakar's population being covered. This demonstrates the importance of waste incineration as both an energy source and waste treatment solution. Dakar can also be served by the proposed Thies incinerators, since the distance between the two cities is only 57 km. Waste can be transported from Dakar to Thies for incineration and conversion into useful energy.

Table 7.16 Financial overview for designing the WtE incineration facility[a]

Exchange rate			
Local currency	CAF		
1 EURO =	655.88		Ref: 2023 (www.xe.com)
Cost estimation of incineration plant (Size: 200,000 t/a)			
Component	Investment %	Cost in €	CAF
Estimated civil engineering investment cost (Part 1)			
Design and engineering, soil reports and permitting, construction management	10%	1.500.000	
Construction costs including construction and general contractor fees	89%	13.350.000	
Other	1%	150.000	
Sum of part 1	**100,00%**	**15.000.000**	
Investment costs on material and equipment (Part 2)			
Waste handling, crane, crane grate incinerator grate (furnace) and boiler (economizer, evaporator and superheater)	52%	52.000.000	34.105.760.000
Power generation (generator, steam turbine, condenser, pumps….etc.)	16%	16.000.000	10.494.080.000
Flue gas treatment (DeNox, Sulfur reduction, HCL, HF, Heavy metals, fly ash)	17%	17.000.000	11.149.960.000
Mobile equipment contingency (10%)	15%	15.000.000	9.838.200.000
Sum of part 2	**100.00%**	**100.000.000**	**65.588.000.000**
Total setup costs		**115.000.000**	**75.426.200.000**
Investment cost management			
State grants	30%	34.500.000	22.627.860.000
Own contribution	0%	2.00	1.312
Additional investment required (loan)	70%	80.500.000	52.798.340.000
Yearly costs for operation and maintenance			
Personal cost	25%	2.500.000	1.639.700.000
Facility maintenance	15%	1.500.000	983.820.000
Water		–	–
Consumables, analyses etc	10%	1.000.000	
Residuals haul and disposal	15%	1.500.000	983.820.000
Contingency (10%)	5%	500.000	327.940.000

(continued)

Table 7.16 (continued)

Total operation and maintenance cost/year	**7.000.000**	**4.591.160.000**	
Yearly costs for capital, consumption, and operation			
Capital related cost including insurance		3.000.000	1.967.640.000
Operation and maintenance related cost		7.000.000	4.591.160.000
Total yearly cost (€/year)	**10.000.000**	**6.558.800.000**	
Technical data			
Parameter	**Unit**	**Value**	
Waste mass flow loading rate	**kg/s**	**6.34**	
Power produced	**MWe**	**12.4**	
Annual power production	**MWh**	**89,013**	
Annual mass flow of bottom ash (15% of total capacity)	**Ton/year**	**30,000**	
Assumed gate fee (if applicable)	**€/ton**	**40**	
Yearly expected profit from the sales of materials			
	Price	**Revenue €**	
Case 1			
Electricity	5 cent €/kWh	4.450.650	
Separated valuable materials from bottom ash (Fe and Al)	40 €/ton	1.200.000	
Gate fees	40 €/ton	8.000.000	
Total		13.650.650	
Case 2 if the plant is owned by the government			
Electricity	18 cent €/kWh	16.022.340	
Separated valuable materials from bottom ash (Fe and Al)	40 €/ton	1.200.000	
Gate fees	40 €/ton	8.000.000	
Total		25.222.340	
Case 3 Gate fee not applicable			
Electricity	18 cent €/kWh	16.022.340	
Separated valuable materials from bottom ash (Fe and Al)	40 €/ton	1.200.000	
Total		17.222.340	
Amortization Year	***< 15 years (assumed)***		

The capital cost of waste incineration costs 500 €/ton, the capacity of the proposed plant is 200,000 ton/year, total cost will be 100 million € without contingency. Estimated civil engineering investment cost is 15% of the total and Investment costs on material and equipment adds to 85% of the total investment

Table 7.17 Financial overview for designing the WtE Gasification or Pyrolysis facility[a]

Exchange rate				
Local currency	**CAF**			
1 EURO =	**655.88**		ref: 2023 (www.xe.com/)	
Cost estimation of incineration plant (Size: 780 t/a)				*Notes*
Component	Investment %	Cost in €	CAF	
Estimated civil engineering investment cost (Part 1)				
design and engineering, soil reports and permitting, and construction management	30%	83.958		
Construction costs include construction and general contractor fees	70%	195.902		
Other		0		
Sum of part 1	**100.00%**	**279.860**		
Investment costs on material and equipment (Part 2)				
Mixer	1%	50.000	32.794.000	
Pumps and headers	6%	220.000	144.293.600	
Heater (heat exchanger)	21%	850.000	557.498.000	
Gasification reactor	17%	670.000	439.439.600	
Cooler (heat exchanger)	1%	27.000	17.708.760	
Valves	1%	21.000	13.773.480	
Liquid gas separator	35%	1.400.000	918.232.000	
Pressure swing adsorption	19%	760.000	498.468.800	
Sum of part 2	**100.00%**	**3.998.000**	**2.622.208.240**	
Total setup costs		**4.277.860**	**2.805.762.817**	
Investment cost management				
State grants	30%	1.283.358	841.728.845	
Own contribution	0%	2.00	1.312	
Additional investment required (loan)	70%	2.994.502	1.964.033.972	
Yearly costs for operation and maintenance				
Personnel cost (labor cost)		101.500	66.571.820	14,500 €/year per operator, a total of 7 operators required
Direct supervisory and clerical labor cost		10.150	6.657.182	10% of operating labor cost
Water		62.400	40.926.912	0.02 €/kg

(continued)

Table 7.17 (continued)

Exchange rate				
Maintenance and repair costs		213.893	140.288.141	5% of fixed capital investment
Operating supplies		21.389	14.028.814	10% of maintenance and repair cost
Laboratory charges		10.150	6.657.182	10% operating laboratory cost
Plant-overhead		162.772	106.758.571	50% operating laboratory, supervision, and maintenance costs
Administrative costs		48.831	32.027.571	30% plant-overhead costs
Additional expenses (e.g. marketing, logistics, operations services, etc.)		42.779	28.057.628	1% of fixed capital investment
Total operation and maintenance cost/year		**673.864**	**441.973.822**	
Yearly costs for capital, consumption and operation				
Capital related cost including insurance		–	–	
Operation and maintenance related cost		673.864	441.973.822	
Total yearly cost (€/year)		**673.864**	**441.973.822**	

Estimated of the equipment investment cost is 7% of the total investment cost

In Touba, two incinerators were proposed which could supply 650,000 inhabitants with power. In Kaolack also two incinerators were proposed, supplying 650,000 inhabitants with electricity according to the power consumption rate/capita as can be seen in Table 7.15. It can be concluded that seven incinerators treat around 1.4 million tons/year of waste which represent 50% of the annual amount of waste generated in Senegal.

Waste-to-energy systems are directly linked to industrial operations, as many industrial facilities require thermal energy or power for their production processes. Therefore, Materials Recovery Facilities (MRF) and energy recovery plants need to be located close to industrial areas. In Senegal, the industrial sector consists of major sectors including food and beverages, textiles, construction and building materials (cement), chemicals, refineries, and mining. Construction and building materials have the highest industrial share in Senegal at 11.7%, followed by mining at 1.9%. Most of these industries are concentrated in the Dakar region.

Fig. 7.9 Waste to energy, MRF potential sites, and existing industries in Senegal

For cities with smaller populations that are located far from the proposed incinerators, existing small incinerators attached to hospitals can be utilized. These facilities are currently used to treat medical and hazardous waste generated by the hospitals themselves. The flue gas cleaning systems for such facilities can be upgraded and developed to handle both hazardous waste from hospitals and municipal waste from villages or cities.

Approximately 31 hospitals exist in Senegal, meaning that 31 small incinerators are attached to these facilities. Due to population distribution patterns, these hospitals are primarily located in the Dakar region and Thies to serve the densely populated cities. More than 5 hospitals are located in Dakar alone. In the southern part of the country, around 5 hospitals are found in regions such as Kolda, Kedougou, and Sédhiou, while in the northern part of the country, only two hospitals exist, with one located in the Saint-Louis Region, as shown in Fig. 7.9.

Generation of RDF (Refuse Derived Fuel)

Refuse-derived fuels are used in a wide range of specialized waste-to-energy facilities, which are using processed refuse-derived fuels with lower calorific values of 8–14 MJ/kg in grain sizes of up to 500 mm to produce electricity and thermal energy (heat/steam) for district heating systems or industrial uses like in the cement industry. In Senegal, cement industry and building materials are existing mainly in Dangote, Sococim and du Sahel which are in Dakar or close to the Dakar region (see Fig. 7.9). Due to that four MRF facilities have been proposed, with two of them serving Dakar region and other regions close to Dakar like Thies and Kaolack. Each MRF facility accommodates 30,000 TPA of mixed waste which can generate around 5000–10,000 TPA of RDF that can be used as a fuel for the cement industry. The other two facilities are in Tambacounda and Matam. The total amount of RDF can be generated is 20,000–40,000 TPA. In addition, RDF also can be used to enhance the energy content of the incinerated waste specifically during the rainy seasons where the lower calorific value of the waste will be lowered.

Gasification/Pyrolysis

Gasification and pyrolysis are considered advanced technologies for waste treatment and energy conversion. Both technologies require specific pretreatment steps before waste can be loaded into the plant. As mentioned previously, the capacities are limited compared to mass burning technology, with mass loading rates of 10–100 tons/day for pyrolysis and 250–500 tons/day for gasification respectively.

In Senegal, mixed waste is generated and the country still has an underdeveloped waste management system. Consequently, both technologies have been proposed for small-scale implementation in Laminia (Kédougou region) and Ranerou (Matam region) respectively (see Fig. 7.9), where waste generation quantities are very low. The energy recovered in the form of synthesis gases can be used immediately or stored for later use in the mining industry.

Furthermore, these plants can be utilized as reference facilities for research and development by universities to optimize processes and enhance energy generation. It should be noted that gasification technology can be incorporated into hydrogen generation as a pilot project, specifically for research purposes and capacity building.

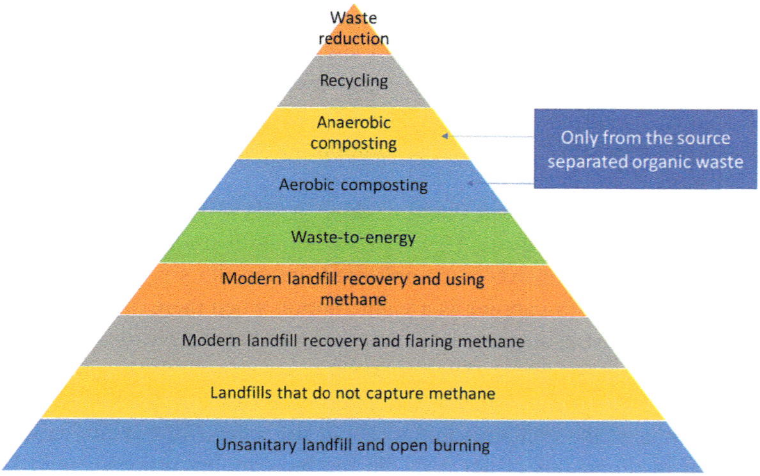

Fig. 7.10 Hierarchy of Sustainable Waste Management at landfill[22]

7.7 Centralized Sanitary Landfill

The Earth Engineering Centre at Columbia University has developed the Hierarchy of Sustainable Waste Management (Fig. 7.10), which is commonly used as a guide for sustainable solid waste management and disposal. In the particular context of study, the practice of "Unsanitary Land filling and Open Burning" has been included in the original waste management hierarchy, which ultimately ends with sanitary landfills (SLFs). The unsanitary practice of landfilling and open burning involves the indiscriminate disposal and incineration of Municipal Solid Waste, which reflects the current state of Solid Waste Management in Senegal and other developing countries.

The waste management hierarchy recognizes that minimizing material consumption and repurposing materials are the most ecologically sound methods. Source reduction entails the act of minimizing the production of waste and reutilizing materials to prevent their entry into the waste stream, thereby postponing waste generation until the stage of reuse. Proper collection of waste is crucial as it is generated. The utilization of recycling and composting is widely acknowledged as the most efficient approach to waste management. Although there are limitations in terms of technology and finances for recycling, as well as difficulties related to product design and inadequate separation of waste at its source, it is imperative to prioritize the promotion of source separation at the local level. Concurrently, alongside improving recycling endeavors, measures should be implemented to handle non-recyclable waste that is currently being produced and will continue to be produced in the future. An effective approach to managing non-recyclable waste

[22] Dutta (2015).

entails harnessing energy through recovery processes. Material recovery is prioritized over energy recovery from waste. Landfilling municipal solid waste (MSW) is equivalent to interring potential secondary resources or sources of energy. While landfills serve the purpose in disposing of a small portion of waste, unsanitary landfilling or open dumping unacceptable solution for managing municipal solid waste and is strongly discouraged.

Reclamation of the dumpsite could be done in two ways. The first method would be bio-capping of the dumpsite. Usually bio-capping is used where the reclamation of legacy waste is expensive or complicated due to the size and the quantity of accumulated waste. Bio-capping is a viable option when there is a lack of technical infrastructure, financial investment or market for materials from the mining of a large volume of legacy waste. The dump sites are capped by placing a cover over contaminated waste to prevent contact between the waste and the environment. This method eliminates the contaminants to avoid the spread of contamination. When reclamation of dumpsites is not possible the bio-capping process could be set to transform dumpsites into a natural environment such as woodlands or parks with a rider condition of post-closure monitoring. The concept of capping is the same as landfill bottom liner.[23]

The conventional cover system consists of four layers of vertical percolation (topsoil), lateral drainage layer (sand), geo-membrane layer and barrier soil layer (clay) placed on the top of the waste to isolate the waste, minimize infiltration and control fugitive gas emission. Another capping option is a water balance (evapotranspiration cover). Evapotranspiration cover system is a natural approach to landfill closure and requires medium compaction effort. Here the main idea is the cultivation of a vegetation layer to control the evaporation from the dumpsite. For this reason the top soil layer should have enough capacity for the plant's roots.

The second method is bioremediation or bio-mining and during the process of bio-mining soil present in the dumpsite is also recovered along with the available recyclable materials. Mining of a dumpsite requires comprehensive information on the properties of waste, site-specific parameters (type, location and land price, volume, composition) project-specific parameters (Stakeholder perspective, choice of technology, the project set-up, licenses, etc.) systemic context (Legislation, markets, regional infrastructure, etc.) and timing of the mining (Future development of costs, prices, legislation, available data and information). Decisive aspects in the planning phase of biomining include:

- Composition of the landfilled waste (waste characterization)
- State of decomposition
- Risks and concerns
- Different waste processing techniques
- Production rates achievable with different techniques
- Soil and moisture content
- The extent of the hazardous materials and landfill gas

---

[23] Presented by Frank E. Pugsley in the processing of the ISWA-SWIS winter school (summer edition) on sustainable landfill management.

Table 7.18 Cost and revenue of the project

Costs	Revenues
Site preparation	Reversal of financial provision
Rental or purchase of equipment	Reversal of landfill tax
Construction or expansion of MRF	Electricity and heat
Rental of purchase of hauling equipment	Selling of the reclaimed materials
Labor cost	Land value
Fuel and maintenance	Energy subsidies
Administrative, planning and regulatory compliance expenses	Avoided post closure monitoring
Worker training in safety procedures	Avoided contamination
Hauling costs	
Transportation	
Re-landfilling, incineration etc.	

- Operational issues (i.e. odour, dust, noise problems, leachate, and stormwater management)
- Public perception
- Economic feasibility.

As a primary step in biomining the excavated waste from the dumpsite is heaped into a large pile to make a windrow of the excavated waste. The primary step is followed by the stabilization of the waste by the exposure of all the waste to air combined with the use of bio-cultures that fosters the composting of any organic matter. This stabilized waste is then screened to recover all valuable products and rejects (organic fines and recyclables). The next step is the management of recovered products through recycling, co-processing and the utilization of other waste management options.

In most of the cases the land area of the dumpsite reclaimed by the process of bioremediation can be used for other alternative purposes. Therefore, for each dumpsite a feasibility study is required before biomining to consider cost and revenue of the project as shown below (Table 7.18 and Fig. 7.11).

7.7.1 Overview of the Size of the Region and Zone/ Region-Wise Centralized Sanitary Landfill in Senegal

Finding a landfill site location is a crucial task in planning a sustainable solid waste treatment and disposal system. Environmental, geotechnical, economic and socio-economic factors should be considered for locating a new site (see Table 7.19). To this end, first the projects and their needs should be defined. Major environmental factors such as airport safety, flood plains, wetlands, fault areas, seismic zones and

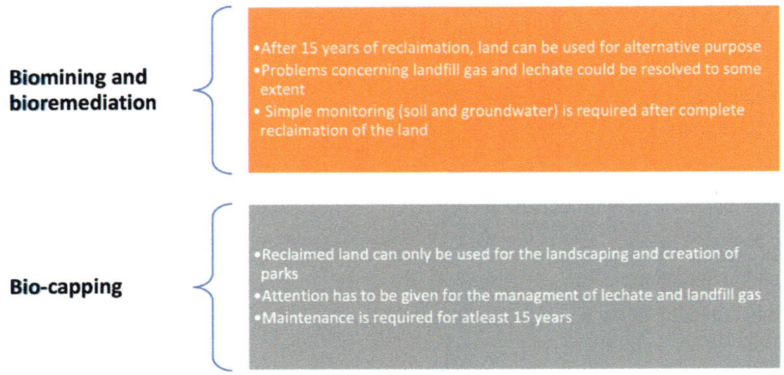

Fig. 7.11 Comparison between bio-capping and bio-mining

the unstable area should be taken into consideration. Accordingly to all the specified requirements and its suitability, candidate sites have to be identified. Environmental, economic and socio-economic data for all the chosen locations are analysed to evaluate and compare candidate sites.

The area of the landfill is another important factor to compare candidates and choose a location. Availability of land for the development of a landfill to receive municipal waste for at least 20 years should be assured. Thus to estimate waste

Table 7.19 Factors to be considered during site selection of sanitary landfill

Environmental/geotechnical	Economic	Socio-economic
Aesthetic impact	Access to highways and available highway capacity	Archaeological and historical sites
Agricultural preservation area	Compatibility with existing solid waste management systems	Cultural patterns
Air quality, gas compositions, and particular matter	Cost of police, fire, and road maintenance	Dedicated land
Areas with high groundwater level	Development, operation, and maintenance cost	Economic and community resources
Climate and atmospheric conditions	Distance to waste generation locations	Emergency responses
Distance from water supply wells	Economic effects	Land use and zoning
Fault area	Effect on property value	Noise impact
Flood plains and wetlands	Land development	Proximity to school and residents
Forest, wilderness, and Scenic area	Highly productive agricultural area	Public safety and health
Geology	Flexibility	Sensitive receptors

generation during landfill lifetime the annual population growth is calculated using the formula:

$$Pf = P_{\mathrm{i}}(1 + r)^n$$

where, Pf = Projected population, P_{i} = Current population, r = Growth rate, n = period of time

The estimated waste generation is used to calculate the volume of receiving waste and required land area as follows:

$$V = \left[\frac{R}{D}\left(1 - \frac{P}{100}\right)\right]Cv$$

where, V = Volume of waste generated, R = Solid waste generation after recycling (lb/person/year), D = Un-compacted density of solid waste (lb/yd^3), P = Volume reduction, Cv = Cover soil ratio

The area required for the creation of the landfill is calculated using,

$$A = \frac{27 \times V \times N}{d(43560)}$$

where, A = Required area for landfill, d = Depth of the landfill (ft), V = Volume of waste generated (cubic yard/person/day), N = Number of People

The total area of the landfill is the sum of the required area annually. Following the above calculation the area required for the development of a new landfill in each district/region of Senegal is calculated by considering the 3% population growth, 0.80 kg/person/da which is the waste generation rate in Dakar, 0.57 kg/person/day the waste generation rate in rest of Senegal, 20 lb/ft^3 Un-compacted density of solid waste, 30% volume reduction, 1.2 cover soil ratio, 50 ft depth of landfill and 20 years life span.

Figures 7.12, 7.13 and Table 7.20 depicts the area required for landfilling 40, 20 and 10% of the total waste generated waste in the various regions of Senegal considering the waste deposition from 2023 till 2051.

7.7.2 *Financial Overview for Designing the Landfill Facility*

Table 7.21 provides information regarding the design, building and operation of a sanitary landfill with a capacity of three thousand tons of waste being received each day. According to the preliminary calculation the construction expenses for a sanitary landfill with a capacity of 3000 t/day are expected to total around 5.625 million Euros. Even taking into account the cash generated by the tipping charge of 225,000.00 Euros, this sanitary landfill requires 310,000.00 Euros/year for its operation and maintenance which should be borne by the local municipality or the government. The

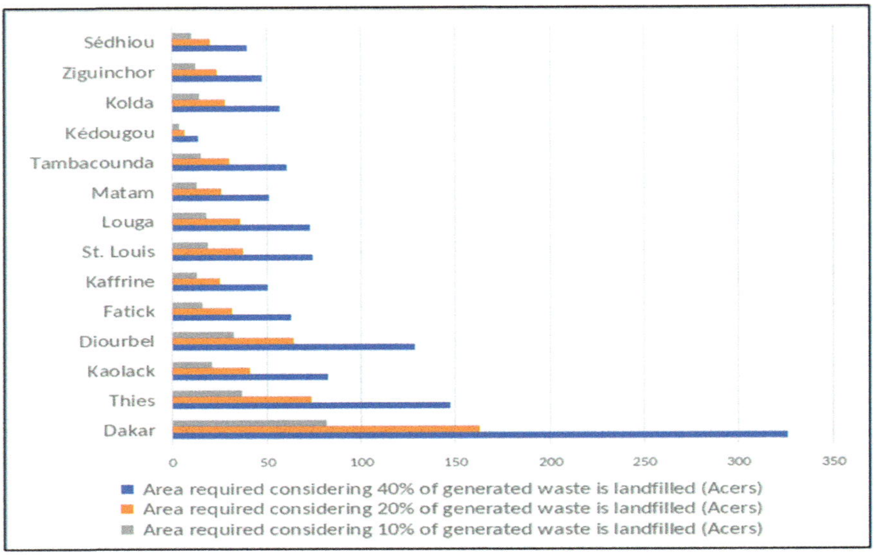

Fig. 7.12 Sanitary landfill area required by various regions of Senegal till 2050 (area in Acres)

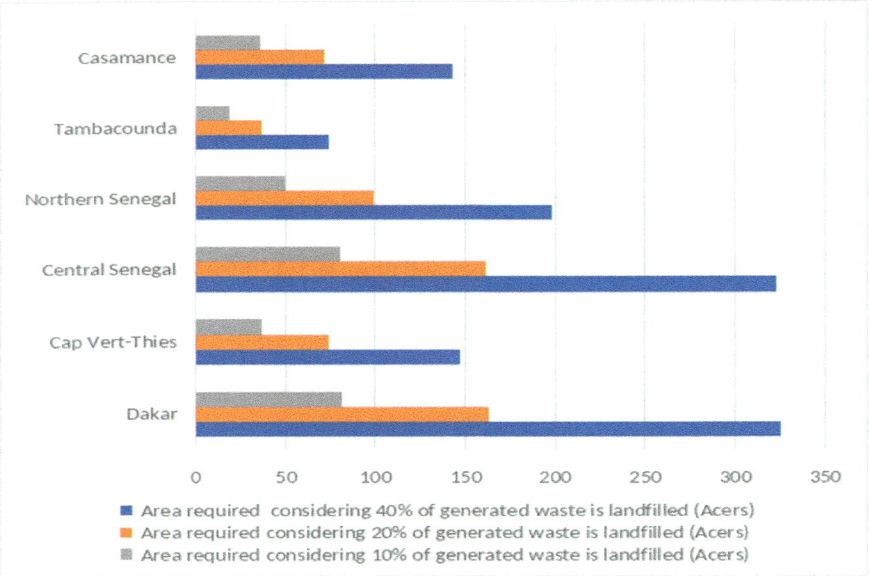

Fig. 7.13 Sanitary landfill area required in various zones of Senegal till 2050 (area in Acres)

Table 7.20 Sanitary landfill area required by various regions of Senegal until 2050

Zone	Region	Population in 2050 by region	Area required f considering 40% of generated waste is landfilled (Acres)	Area required considering 20% of generated waste is landfilled (Acres)	Area required considering 10% of generated waste is landfilled (Acres)
Dakar	Dakar	8,978,949.95	325.5	162.75	81.375
Cap Vert-Thies	Thies	5,065,556.28	147.2	73.6	36.8
Central Senegal	Kaolack	2,814,404.27	81.79	40.895	20.4475
	Diourbel	4,399,975.91	127.866	63.933	31.9665
	Fatick	2,145,569.71	62.351	31.1755	15.58775
	Kaffrine	1,737,654.41	50.497	25.2485	12.62425
Total area in acres			322.504	161.252	80.626
Northern Senegal	St. Louis	2,554,620.08	74.23	37.115	18.5575
	Louga	2,491,439.95	72.4	36.2	18.1
	Matam	1,753,083.49	50.945	25.4725	12.73625
Total area in acres			197.575	98.7875	49.39375
Tambacounda	Tambacounda	2,081,752.07	60.497	30.2485	15.12425
	Kédougou	452,527.66	13.151	6.5755	3.28775
Total area in acres			73.648	36.824	18.412
Casamance	Kolda	1,945,093.93	56.525	28.2625	14.13125
	Ziguinchor	1,621,432.13	47.12	23.56	11.78
	Sédhiou	1,360,870.49	39.548	19.774	9.887
Total area in acres			143.193	71.5965	35.79825

plant requires land that is conveniently located within close proximity to major road networks. It is also important to note that the cost information will vary depending on whether the site in consideration is an existing and operational landfill or a proposed site as well as the degree to which the planning, resource consent processes and design have progressed. The prices shown are for purposes of illustration only and are subject to change based on a variety of factors including those related to the local market and economy as well as those particular to the site-specific data.

Table 7.21 Financial overview for designing the landfill facility

Exchange rate			
Local currency	**CFA**		Ref: 2023 (www.xe.com)
1 EURO =	**655.88**		
Cost estimation of sanitary landfill (Size: 3.000 t/d)			
Component	Investment %	Cost in €	CFA
Typical planning and pre-development costs (part 1)			
Project management, site selection and consultation	2%	100.000	65.588.000
Survey, design and other detailed studies	2%	100.000	65.588.000
Geotechnical and ground water investigation	0%	20.000	13.117.600
Land acquisition and associated/set-up costs	1%	30.000	19.676.400
Sum of part 1	**4.44%**	**250.000**	163.970.000
Typical construction and development costs (part 2)			
Site access—intersection upgrade, other roading network upgrades, and special structures (diversions, bridges, etc.)	26.67%	1.500.000	983.820.000
Site amenities and services—Site entrance, administration building, weighbridge and kiosk, machinery shed, maintenance facility, sewerage, water supply, general civil works (sealing, parking, etc.), washdown facility/wheel-wash, fencing, landscaping etc.,	22.22%	1.250.000	819.850.000
Cell construction—Sediment control structures and measures, groundwater control/subsoil drainage, leachate collection, transmission, and pre-treatment, leachate disposal system	14.22%	800.000	524.704.000
Storm water management system	16.00%	900.000	590.292.000
Gas management system—Horizontal collectors, vertical extraction wells, ring header (below grade), laterals to vertical wells (above grade), condensate traps, flare stations, etc	13.33%	750.000	491.910.000
Final cover system—Geosynthetic layer, drainage layer, erosion control (vegetation)	1.78%	100.000	65.588.000

(continued)

Table 7.21 (continued)

Office equipment and small tools	1.33%	75.000	49.191.000
Sum of part 2	**95.56%**	**5.375.000**	3.525.355.000
Total setup costs	**100.00%**	**5.625.000**	3.689.325.000
Investment cost management			
State grants	30%	1.687.500	1.106.797.500
Own contribution	0%	0	0
Additional investment required	70%	3.937.500	2.582.527.500
Yearly costs for operation and maintenance			
Salaries, wages and overhead including: on-site management, gate control and fee collection, audit fees, secretarial fees, accounting fees, legal, consultants, insurance, waste acceptance and inspection, and health and safety		300.000	196.764.000
Energy demand (at 0.5 kWh/ton)		45.000	29.514.600
Fuel demand (at 0.1 L/ton)		90.000	–
Leachate treatment and disposal		20.000	13.117.600
Stormwater maintenance		20.000	13.117.600
Monitoring–stormwater, groundwater, leachate, landfill gas		20.000	13.117.600
Machine maintenance		30.000	19.676.400
Insurance		10.000	6.558.800
Total operation and maintenance cost/a (gross)		**535.000**	350.895.800
Yearly costs for capital, consumption, and operation			
Capital related cost (interest on additional investment at 6% p.a)		236.250	154.951.650
Operation and maintenance related cost		535.000	350.895.800
Total yearly cost (€/year)	**771.250**	505.847.450	
Yearly profit from tipping fee			
Tipping fee (at 0.25 €/ton)		225.000	19.125.000
Yearly costs for operation and maintenance after profit			
Net yearly costs		310.000	26.350.000

7.7.3 Approach and Timelines of Implementing the Sanitary Landfill Creation and Management

The action plan presents the achievable goals, approach and timeline to reach the set goals in a set timeframe. These targets and action plans will guide the state's efforts in the creation of Sanitary landfills, cleaning up the legacy waste of the existing old dumpsites which are not receiving any more waste and those which have reached their full capacity. As an estimation, management of existing dumpsites and the creation of new sanitary landfills could create about jobs by the time the sanitary landfills become operational in Senegal. Priority should be given to the closing of small dumpsites and the creation of sanitary landfills construction at hotspots across Senegal.

As a **short-term goal** (up to 1 year) identification of hotspots for the creation of centralised sanitary landfill and the creation of the legislation for the closing of dumpsites which are small and non-functional and partial reclamation of closed dumpsites has to be prioritised. By the end of the first year targets should be set for closing all the dumpsites (in a phased manner) in Senegal which are outside the specified hotspot locations and importance has to be given to receiving the waste only in the hotspot landfill sites. In addition to this the reclamation has to be started on at least 50% of the closed dumpsites and efforts have to be made to fit the leachate and landfill capture system in at least 30% of the hotspot landfill sites.

The **medium-term goals** (2nd and 3rd year) start from the second year of action plan implementation and continue until the end of the third year. During the first term of the medium-term goal reclamation of at least 75% of the closed dumpsites has to be achieved and a minimum 60% of the hotspot landfill sites have to be fitted with the leachate and landfill capture system. Further during the second term of the medium-term goal all the closed landfills have to be completely reclaimed using appropriate and 100% of the hotspot landfill sites have to be fitted with the leachate and landfill capture system. It is important to integrate the MRFs into the hotspot landfill sites within the geographical vicinity and proper legislation has to be made to accept only stabilised waste into the landfill sites.

As a **long-term goal** (4th year and beyond) processing of waste at all treatment facilities has to be emphasized and the regional sanitary landfill sites should accept only stabilised waste from region of location. When requirement for additional landfill arises, permission has to be granted only for the construction of engineered sanitary landfills after proper evaluation of the site of construction. Adaptation of mentioned strategies not only aids in effective management of landfill sites but also has huge potential to mitigate the damage to the environment including soil, water and air pollution caused by unsanitary dumping of waste in Senegal. Figure 7.14 gives an overview of action plans and approach timeline for the creation of a sanitary landfill.

This strategy on the transition from dumpsites to sanitary landfills and suggestions on hotspot creation have their specific designated benefits to enhance the waste management system of Senegal. Nevertheless the implementation of this strategy will also have an interlinking effect between the strategies that enable the management of waste in Senegal in the broader context.

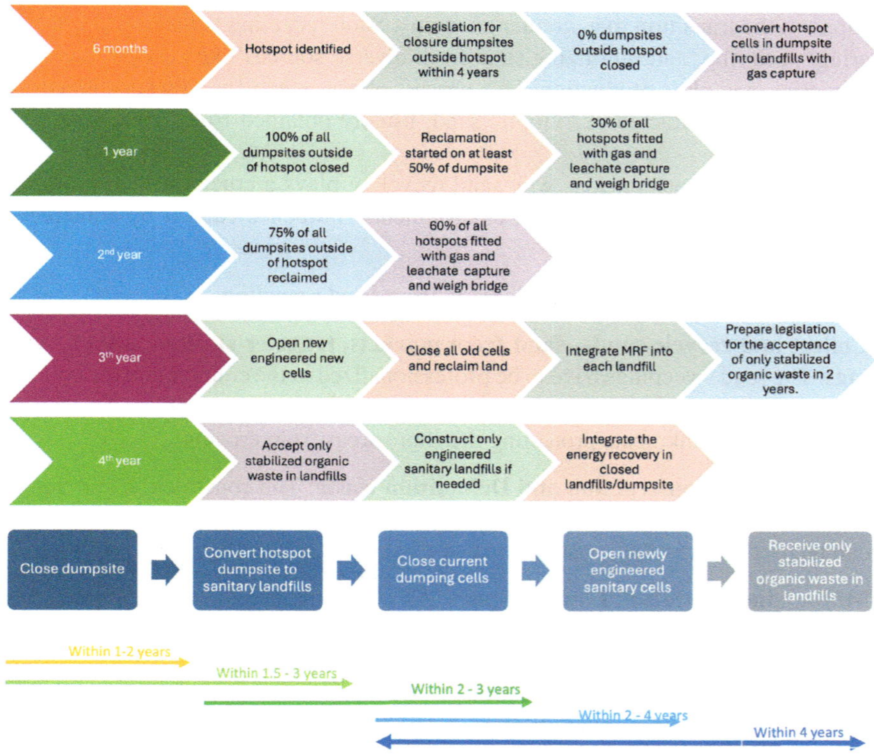

Fig. 7.14 Time-line and action plan for the existing and proposed landfills

7.8 Recycling of Sand and Fine Element Waste

Sand and fine element make up the majority of Senegal solid waste which contributes to the difficulty and expense of the country waste management. The composition of solid waste in Senegal varies from large urbanized areas like Dakar to secondary cities according to a study that was carried out nationwide by the Unite de Coordination et de Gestion (UCG) but in general it is made up of two thirds of fine particles, sand and organic waste. The sand coming from sweeping the communal streets and houses yard should be collected separately into a define storage bin (identified via colour or inscription). The storage sand bins can be placed at location accessible to households and communal waste workers filled sand bins will then be transported to be sieved and filtered sand portion is disposed of along the sea or water-bodies protecting the shore. The debris from the sieving will the then be transported and disposed of at the landfill.

As we strive to create a more sustainable future it's essential to explore innovative ways to recycle and reuse materials that would otherwise go to waste. One such method is sand recycling which offers a multitude of environmental benefits and supports the transition to a circular economy. Sand recycling is the method of gathering, purifying, and reutilizing sand from diverse sources like construction areas, foundries, and glass production. This process offers an eco-friendly substitute to the use of freshly extracted or "virgin" sand, usually obtained from rivers, lakes, and ocean beds.

There are various reasons why sand recycling plays a crucial role in promoting sustainability and decreasing the environmental influence caused by construction actions and are listed below:

Economic and Environmental Benefits from sand recycling

Collect and recycle sand—sand from markets, transfer stations, street sweepings: The road sweepings have to be professionalized in the country and the collected sand can be recycled to use in construction. Further this treated sand can be also used in creating embankments protecting the shore areas from erosion.

Treatment of Construction and Demolition waste: The construction and demolition waste has to be processed and can be used filling either the low lying areas or can be used back in the construction.

Minimizing Waste and Landfill Space: As we continue to develop our urban areas it's important that we take steps to minimize our impact on the environment. One way we can do this is by utilizing sand recycling to reduce the amount of waste produced from construction and demolition projects. By doing so we can help conserve landfill space and reduce the amount of waste that ends up in landfills. It's an important step toareas a more sustainable future.

Reducing Resource Depletion: Given the depletion of natural resources, the demand for sand across different industries, particularly construction, is rapidly increasing. Recycling sand presents a practical and sustainable solution to conserve these resources for future generations and minimize the need for extracting virgin sand.

Lowering Greenhouse Gas Emissions: Choosing to recycle sand instead of extracting, transporting, and processing new sand is a crucial step toareas reducing greenhouse gas emissions and supporting global efforts to combat climate change. The production of virgin sand generates a significant amount of emissions, making it imperative to consider environmentally friendly alternatives like sand recycling.

Supporting a Circular Economy: The concept of a circular economy strives to reduce waste and maximize the utilization of resources by prolonging the lifespan of materials. An exemplary instance of this idea is the process of sand recycling, which significantly contributes to sustainability and resource efficacy in multiple industries.

Sand recycling production facility and approach

The process of recycling the fine element and sand fraction in municipal solid waste (MSW) involves several steps that aim to effectively clean, process and reuse. Here's a detailed description of the process:

Collection: Sand is collected from communal street sweepings with sweeping machine, construction and demolition sites, foundries, or other sources.

Sorting/Screening: A simple and cost-effective scheme is suggested for the sorting plant. The sand waste collected is guaranteed to be pure through a thorough cleaning process that eliminates impurities like dirt, clay, and organic matter. The sorting line includes a bag opener, a conveyor belt with a hopper, trommel sieves. Waste bags are entered into the bag opener followed by the trommel sieves to separate the fine fraction. The over-sieve fraction contains recyclable and non-recyclable fractions, whereas the fine fraction mainly consists of sand. The trommel sieves should of different sieve size openings to directly grade the sieved materials and ease the process.

Grading: This process is required to refine the final product. Rotating drum sieve is suggested here. Based on the final application and local condition various mesh sizes have to be used. After the sand has been sieved, it undergoes a grading process based on its size and composition. This is done to ensure that it meets the required industry standards and the reuse purpose of sand for the protection of embankments of various water bodies such as rivers, lakes, and ocean beds.

Bagging and Reuse: Recycled sand is utilized in various applications such as river and ocean beds or as to replace seashore sand including the production of concrete, road construction, and even 3D printing.

The phase of the strategy to be implemented are suggestive and described in Fig. 7.15. For the collection and the recycling of sand waste to succeed it is crucial to prioritize the separation of street sweeping through a new collection scheme in the first year. During the mid-term phase street waste must be gathered using collection equipment like street sweeping machines or mini trucks and taken directly to a sand recycling facility. A recycling facility should also be built to recover sand from the collected waste which can be reused for river and ocean beds and seashores. In phase 3 technology should be improved to treat construction and demolition waste, which can be valuable for cement industries.

7.9 Reducing in GHG Emissions by Adopting Centralised MSW Treatment Facilities

Reducing greenhouse gas (GHG) emissions through improving waste management systems is an effective strategy to combat climate change. Waste management contributes to GHG emissions primarily through the decomposition of organic waste in landfills and the release of methane a potent greenhouse gas.

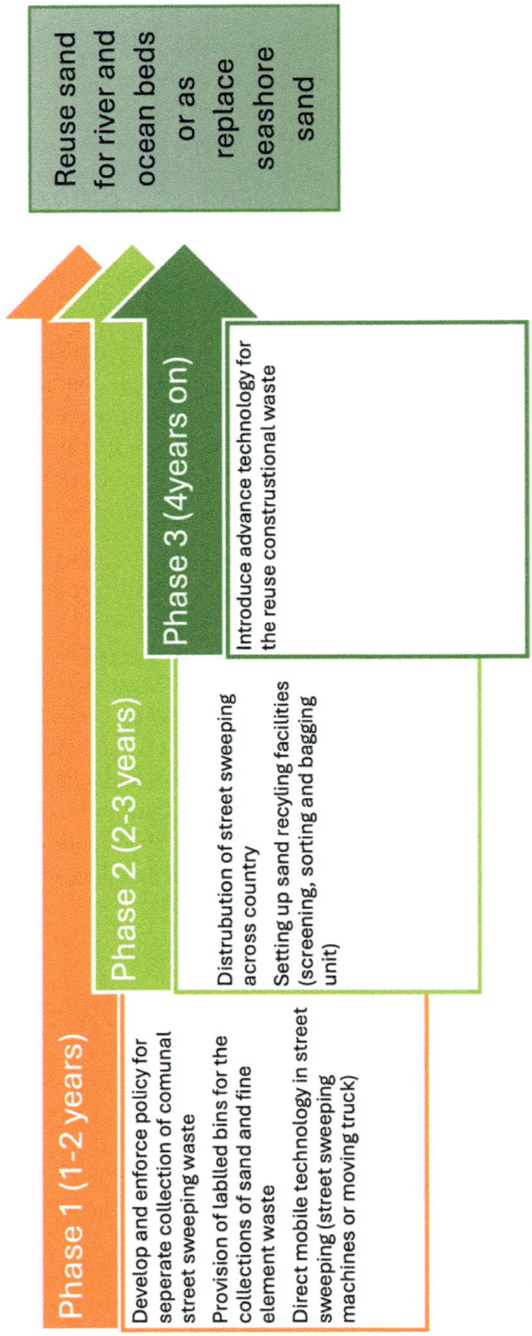

Fig. 7.15 Approach for promoting waste minimization and waste management at source

7.9.1 *Emission Reduction Through Composting*

Composting is a natural process that converts organic waste such as food scraps and other biodegradable materials into nutrient-rich soil amendment called compost. Composting can contribute to the reduction of greenhouse gas (GHG) emissions in several ways:

Methane reduction: When organic waste decomposes in landfills without oxygen, it produces methane, a potent greenhouse gas that contributes to climate change. Composting on the other hand is an aerobic process that occurs in the presence of oxygen which reduces methane emissions. By diverting organic waste from landfills and composting it instead would significantly reduce methane emissions.

Carbon sequestration: Compost contains stable organic matter that helps to sequester carbon in the soil. By adding compost to agricultural fields, gardens or degraded soils enhancing soil organic carbon levels. Increased soil organic carbon not only improves soil health and fertility but also serves as a long-term carbon sink, reducing carbon dioxide (CO_2) emissions in the atmosphere.

Avoided commercial fertilizer application: Compost can replace synthetic fertilizers in agriculture and gardening. The production of synthetic fertilizers is energy-intensive and often relies on fossil fuels contributing to GHG emissions. By using compost as a natural fertilizer we can reduce the demand for synthetic fertilizers and their associated emissions.

Based on the five major zones in Senegal (Dakar, Cap Vert-Theis, Central Senegal, Northern Senegal, Tambacounda, Casamance) the waste which generated mostly lands up in the landfill. As shown in the Table 7.22 and Fig. 7.16, Dakar generates approximately 377,705 TPA of waste which contributes 608 kt CO_2 emissions. In the fraction of the waste that is disposed of in a landfill 12% of the organic waste is compostable.

7.9.1.1 Emission Reduction Through Recycling

In summary municipal solid waste (MSW) composition varies between developed and developing nations but after materials are removed for recycling a residue with a significant calorific value remains. This residue is well-suited for energy recovery operations. On average one tonne of MSW has about one-third of the calorific value of coal with MSW having 8–12 MJ/kg and coal having 25–30 MJ/kg. The energy potential of MSW can generate approximately 600 kWh of electricity. Previously MSW was often incinerated in mass burning facilities but new installations are focusing on higher efficiencies in power and Combined Heat and Power (CHP) production.

As seen in the Table 7.23 when a ton of recovered paper is recycled and converted into new recycled paperboard for packaging and other purposes approximately 3.6 million metric tons of CO_2 emissions are eliminated. This reduction in emissions

Table 7.22 Emissions by total MSW in the landfill, emissions by the organic fractions and reduction of emission by composting the organic fraction

Zone	Possibility of generating CO_2 per tonne of MSW in Kilotonne (kt)	GHG emission by the organic fraction in the landfill in Kilotonne (kt)	GHG emission reduction potential when organic fraction is composted in Kilotonne (kt)
Dakar	608.1	228.0	114.0
Thies	214.4	80.4	40.2
Central Senegal	469.7	176.2	88.1
Northern Senegal	287.8	107.9	54.0
Tambacounda	107.3	40.2	20.1
Casamance	208.6	78.2	39.1

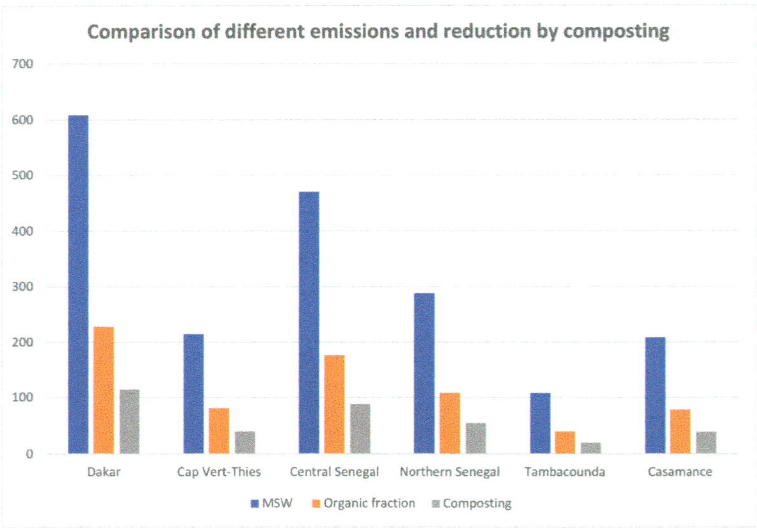

Fig. 7.16 Emissions by total MSW in the landfill, emissions by the organic fractions and reduction of emission by composting the organic fraction

occurs through three key mechanisms. Firstly paper recycling helps prevent methane emissions from landfills. When paper, plastic and mental are not recycled a substantial portion around 80% ends up in landfills. As this paper decomposes in the anaerobic conditions of landfills it releases methane—a potent greenhouse gas that has 21 times the heat-trapping capacity of carbon dioxide (see Fig. 7.17). Landfills are the largest source of methane emissions in the United States and the decomposition of paper is the primary contributor to this methane generation. By recycling paper instead of sending it to landfills we can significantly reduce methane emissions. Similarly

Table 7.23 Emission reduction in tonne of CO_2 per tonne of plastic, paper and glass waste recycled

Zone	Emission reduction (Tonnes of CO_2) by recycling plastic which is 9% fraction in the recyclable waste[a]	Emission reduction (Tonnes of CO_2) by recycling paper which is 5% fraction in the recyclable waste[b]	Emission reduction (Tonnes of CO_2) by recycling glass which is 2% fraction in the recyclable waste[c]
Dakar	637.38	42,492	787
Cap Vert-Thies	224.74	14,983	277
Central Senegal	492.36	32,824	608
Northern Senegal	301.65	20,110	372
Tambacounda	112.44	7496	139
Casamance	218.61	14,574	270
Total emission reduction by the recyclables	1987.17	132,478	2453

[a] https://www.paperrecyclingcoalition.com/policyissues/how-recycling-paper-fights-global-warming/
[b] https://8billiontrees.com/carbon-offsets-credits/carbon-ecological-footprint-calculators/plastic-carbon-footprint
[c] Why Recycle Glass?—Glass Packaging Institute (gpi.org)

plastic and glass recycling also have impact on reducing GHG emissions which do not end up in landfill rather they can be reused. Our estimates suggest that plastic and glass can altogether reduce up to 1,987,17 and 2453 Tonnes of CO_2 equivalent emission respectively.

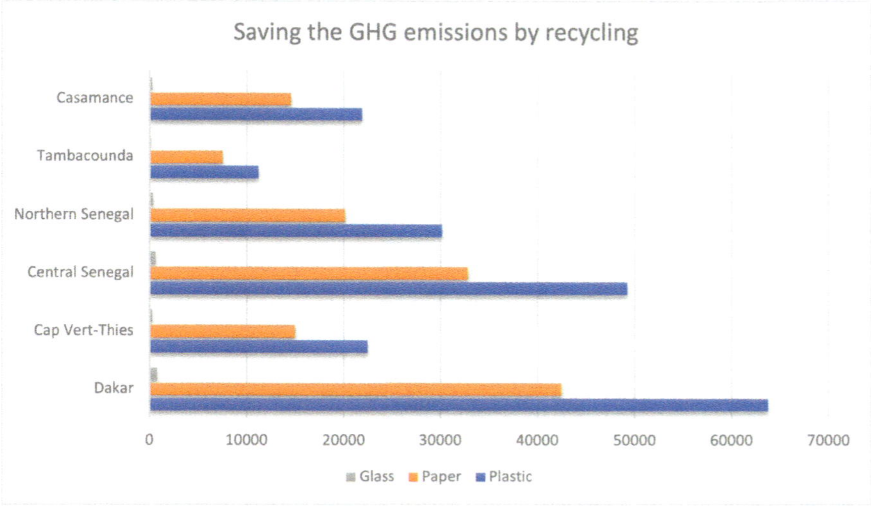

Fig. 7.17 Emission reduction in tonne of CO_2 per tonne of plastic, paper and glass waste recycled

References

Alam et al (2022) Selection of waste to energy technologies for municipal solid waste management—towards achieving sustainable development goals. Sustainability 14:11913. https://doi.org/10.3390/su141911913

Beri KY (2018) Improvement of the waste management system in Senegal. Mediterran J Basic Appl Sci

Cabral et al (2012) Low-level environmental exposure to lead and renal adverse effects: a cross-sectional study in the population of children bordering the Mbeubeuss landfill near Dakar, Senegal. Hum Exp Toxicol

Circular Economy in Africa-EU Cooperation (2020) Country-Report-Senegal_Final_20122020_EN.pdf (trinomics.eu). Retrieved on 16 May 2023

Moyaa D, Aldás C, López G, Kaparaju P (2017) Municipal solid waste as a valuable renewable energy resource: a worldwide opportunity of energy recovery by using waste-to-energy technologies

Dutta AB (2015) Review of solid waste management in India

IDRC (2011) Dakar's Mbeubeuss landfill: More than meets the eye. International Development Research Centre (IDRC)

Kawai K, Tasaki T (2016) Revisiting estimates of municipal soild waste generation per capita and their reliability. J Ater Cycles Waste Manag 18:1–13

Münster M, Lund H (2010) Comparing waste-to-energy technologies by applying energy system analysis

Netherlands Enterprise Agency (RVO) (2022) Sector report circular economy Senegal. Terminomic B.V., Rotterdam

Nojosa E, Barbosa R, Marques G, Vasconcelos O (2021) Biomass reduction of organic materials in a domestic composting system redução de biomassa de materiais orgânicos em sistema de compostagem doméstica. Biofix Sci J 6:98–102. https://doi.org/10.5380/biofix.v6i2.79902

Swachh Bharat Mission Municipal Solid Waste Management Manual (2016) Chrome-extension. https://mohua.gov.in/upload/uploadfiles/files/Part2.pdf. Retrieved on 20 May 2023

Waste Wise Cities Bulletin (2020) wwc_bulletin_5.pdf (unhabitat.org). Retrieved: 16 May 2023

Chapter 8
Strategies for Digitalizing the Waste Management Sector

The rapid increase in population growth, urbanization and economic development has led to an increase in municipal solid waste generation in Senegal. There is a need for proper management of municipal solid waste and if not managed well the waste can become hazardous to human life and the environment in the future. Digital technologies are crucial in an effective waste management sector for a safer, transparent, economically feasible and highly resource-efficient waste management system. Various digital technologies in the waste management sector have been introduced even though the digital field is quite diverse, highly competitive private companies and bigger public players tend to accept and adopt modernized digital technologies faster. Comparatively the smaller companies and public enterprises appear to be slower and lack necessary investments as well.[1,2]

For well-organized monitoring and assessment of waste management, proper Information and Communications Technology (ICT) is required. Figure 8.1 depicts the challenges faced in municipal solid waste management systems without the introduction of ICT. With the help of ICT complaints of people can be analyzed with actual information and solutions can be made accordingly. Wastebins and collecting vehicles can be tracked easily with the help of ICT and the route can be optimized accordingly for more efficient and transparent collection and transportation of waste. Difficulties and complications in calculating the amount of waste generated can be handled well with the help of ICT. Through ICT response to emergencies and proper management of manpower can be achieved easily. User fee collection directly using a digital platform helps in a competent waste management system. Digitalized waste management helps public authorities in providing better services to the people and society and thereby upscaling them one step closer to the concept of a circular economy.

---

[1] ISWA, 2019.

[2] Eionet Report—ETC/WMGE 2020/4: Digital waste Management.

S. Narra et al., *Sustainable Waste Management in Senegal*, World Sustainability Series,
https://doi.org/10.1007/978-3-031-98968-1_8

Fig. 8.1 Challenges in SWM without ICT

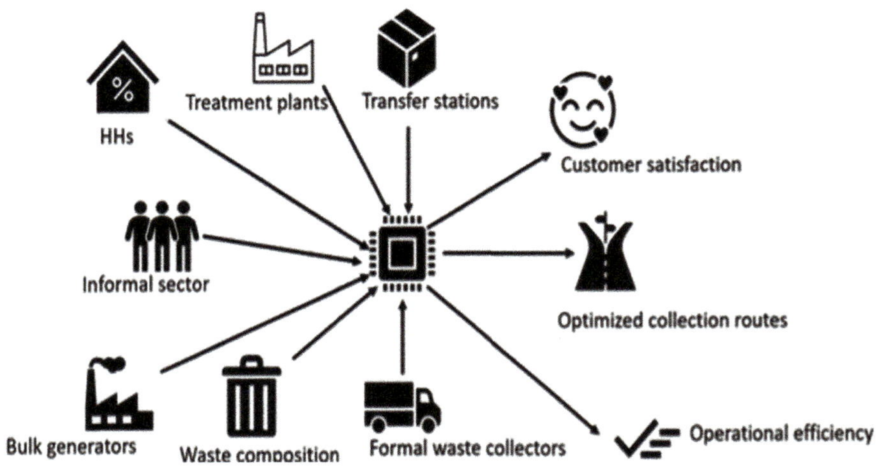

Fig. 8.2 Overview of digitalization in the waste management sector

There are a few factors that delay the use of digital tools in the waste management system. Factors such as digital literacy to implement and monitor digital tools, higher costs for digital solutions, security of IT systems and requirement of digital ecosystems are considered as the inhibition factors in the digitization of the waste management sector. As a result digitalization in the waste management sector of Senegal has to be adopted gradually into all the key sector in the waste management process as identified in Fig. 8.2.

8.1 Communication

Communication has an imperative role in digital waste management systems, especially in tasks such as billing and documentation. Communication between waste collectors and people (customers) about the information on pick up dates, information on waste separation, bills and disposed waste reports are needed to schedule pick-ups, ongoing waste contracts data change and to enquire on prices and services.

Several digital solutions to enhance communication are subdivided into the following categories:

- **Website**: General information on services, prices, fees etc. will be displayed on websites. Information on waste separation, collection timings, data on collection stations, transfer stations, information on hazardous waste and how to deal with it etc. will be mentioned on the websites. It is easy for citizens to report on illegal littering, uncollected or lagged waste collection and other issues through websites as a medium of communication.
- **Mobile App**: Mobile apps are user-friendly these days as they are handy for customers to track real-time data on their waste given away, notifications and messages of important news and events can be received immediately as well.
- **Integration in already existing apps or websites**: if there is already an application or website available for communication purposes waste-related information can be added to this. This can help in minimizing the number of different information access points and thereby increase customer satisfaction and convenience.
- **Usage of third-party apps**: Third-party apps such as WhatsApp, Facebook, Twitter, Instagram etc. can be used to communicate with customers as these apps have better reach among people.

8.2 Waste Collection

The collection of waste is a vital part of waste management and it can be digitized using sensors in collection vehicles with weighing balances to check the weight of the collected waste and route planning to optimize the routes. Thus this can save operation time, inventory tracking and resource planning. Even a minor improvement in routing can have a faster return on investments due to the higher costs for running the collection vehicle. The one possible way to solve main challenges in waste collection, segregation and treatment is by using QR Codes.

Tracking waste collection by QR Codes

To track the waste collection authorities can assign a QR Code to each household. This QR Code will be encoded with a unique identifier or a unique URL. QR Code will make an entry to the online database when scanned by the waste collectors. This can include many details such as time of waste collection, date, household number, amount of waste, information regarding segregation etc. This data can be further processed and analyzed by the authorities. This will ensure that the sanitary workers do not miss collecting waste.

Digitalization of user-fee collection system and alternative benefits

Digitalization plays a significantly greater role in the success of user fee collection. The efficiency of user fee collection can be improved by introducing a digital house-tag system (similar to that of automatic highways toll collection in many countries). In the house-tag system the owner of any residential or commercial will pay a yearly user fee in advance (depending on the designated combination of base, variable and EPR fee) and the difference in the paid fee and the utilized services will be corrected every year and the difference will be settled accordingly. The introduction of such a system in Dakar initially will ensure higher transparency and budget in advance for the smooth functioning of the waste management system.

8.3 Internal Processes

All the paper-based processes such as billing, accounting, controlling, processing of orders, documentation, and management of sub-contractors could be digitalized. The main advantages of digitalization in reducing paper works have cost reduction more storage of data easiness to find data and reduction in chances for errors. The digitalization technology applications in the waste management sector is as shown in Table 8.1.

The digitalization approach and timeline in the waste management sector is as shown in Fig. 8.3. The approach to digitalization is subdivided into 3 stages:

Stage 1 should focus on digitalizing the collection system, transfer system and capturing the dumpsite information. Digitalization at stage 1 will significantly improve the efficient route planning, resource planning, tracking of the vehicles and the movement of waste and enables the systematic collection monitoring and maintenance of all the data. Stage 1 is expected to take up to 8 months to complete.

Stage 2 focuses on digitalizing waste generation such as households, bulk generators, religious places, etc. As explained earlier a QR code has to be assigned to a different

Table 8.1 Examples of digital technology applications in waste management

Communication	Waste collection	Internal processes
Websites	QR codes	Billing
Mobile apps	Sensor-equipped vehicles	Accounting
Integration in other services	Route planning	Controlling
Third-party social media apps	Resource planning	Processing of orders
Third-party social media apps	Inventory tracking	Documentation

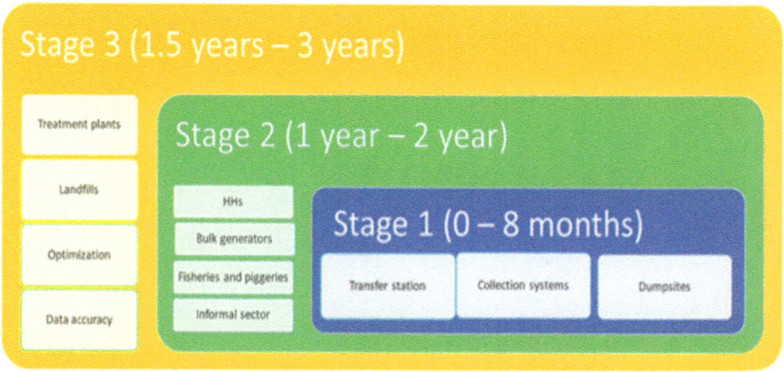

Fig. 8.3 Digitalization approach

type of waste generator using which details such as time and date of waste collection, house number, amount of waste and information regarding segregation could be captured. As a part of stage 2 even the piggeries and animal husbandry also have to be digitalized to have information about the quantity of waste being reused by such a system. This stage is estimated to take a time of 1 year to 2 years to complete the digitalization.

Stage 3 should involve the digitalization of treatment plants, landfills and waste management plants that could enable optimization and data accuracy of the entire waste management system. It is expected to take up to 3 years to complete the digitalization of stage 3.

Strategies enhancement of municipal waste management through stakeholder engagement, awareness-raising and capacity building.

A major component of ensuring the sustainability of waste management systems is the integration of environmental and economic approaches, behavioural change and capacity building activities. Behavioural change requires sustained effort over time which is why these activities need to be considered as a continuous process. It is noted in National Strategy for Integrated and Sustainable Management of Solid Waste that an effective solid waste management system requires the cooperation and input of many stakeholders including groups like community-based organizations, NGOs and other agencies responsible for waste management as well as students. Such stakeholders are encouraged to be involved in the development of interventions that support change in behaviour.

Chapter 9
Strategy on Awareness-Raising Campaign and Communication Channels

Municipalities/UCG/SONAGEDs can adopt a gradual approach to raise awareness, increase public interest in SWM issues, evaluate the community participation, make a trial of different approaches, adapt the best and maintain the approved communication schemes (see Fig. 9.1). In a modern waste management system, people are trained to be aware of methods to handle the waste generated (**Awareness**). To invite people to the voluntary programs e.g. home-composting/biogas and the major focus here would be operation and maintenance. Whereas in the case of obligatory programs e.g. illegal waste dumping it is explained beforehand exactly what is expected. Participation could be increased when program criteria are clear and simple to follow by creating interest in people (**Interest**). Initial participation of the people could be usually fair even in the most well-publicized initiatives (**Evaluation**). People may decide not to participate if they are experiencing trouble (**Trial**). Additional guidance and information should be available through well-publicized helplines. The level of participation should continue to rise (**Adoption**). Ongoing education programs encourage constructive comments and where appropriate it has to be facilitated with updated materials. High participation rates are maintained through encouragement and effective instructions (**Maintenance**).

To raise the awareness of the community education methods are chosen based on the social conditions and budget. People must be informed through several approaches such face to face training, communication via digital apps and social media and regular event at public places for national and international environmental occasions such as earth day, public cleaning campaigns, street plays, billboards, etc. Training of the young generation at school educational institutes etc. is more effective for the future solid waste management system.

The communication chosen in Senegal should be suitable for all the target audiences. In a print medium attention should be paid to the infographic to convey messages noticeably. The icon representing reuse, recycling and disposal can be promoted by artefacts billboards wall writing and printing on all products used by the community. Audio-visual communication has a broad reach and a long-term

S. Narra et al., *Sustainable Waste Management in Senegal*, World Sustainability Series,
https://doi.org/10.1007/978-3-031-98968-1_9

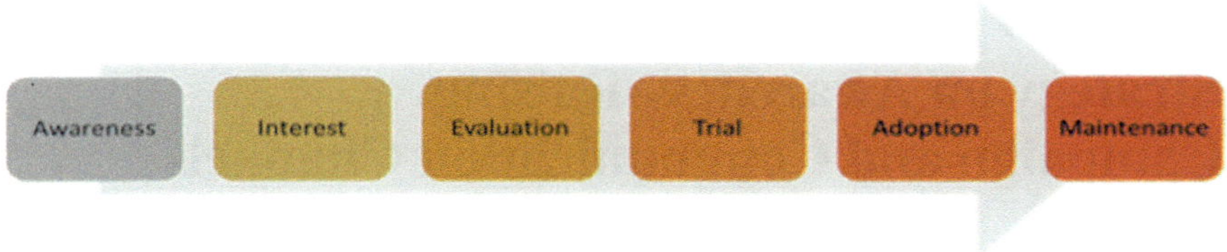

Fig. 9.1 The process of raising awareness through educational programs

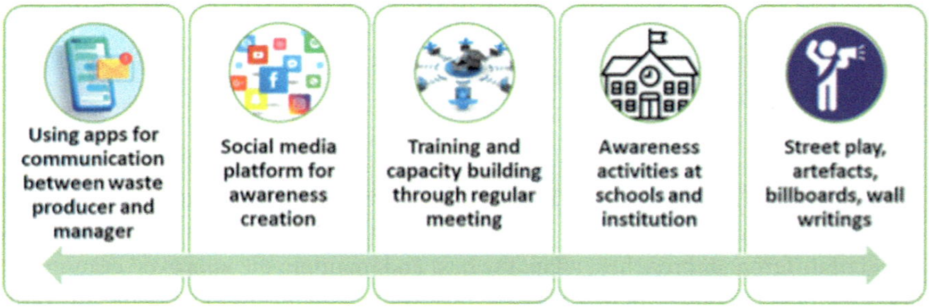

Fig. 9.2 Communication channels could be utilized for the creation of awareness

influence and it is effective in changing attitudes and encouraging behaviour change. Through collaborations with local radio and television stations, a long-term campaign on sanitation and hygiene is achievable. When compared to traditional commercials, catchy advertisement tends to capture the audience's attention and have a stronger recall.

Internet websites and social media could also have an effective reach to the resident community in Senegal. As of now the most successful communication strategy in this aspect is face to face communication which includes persuasive conversations and talks with individual members of the household. Door-to-door collection staff have shown to be the most crucial connection in maintaining source segregation initiatives. Through the awareness raising campaign, all residences, business entities and other institutional establishments must be contacted. Counter parties such as religious leaders, resident welfare associations, community-based organizations, non-governmental organizations and self-help groups can promote awareness-raising through regular meetings. The overview of different communication channels that could be used for creating awareness is provided in Fig. 9.2.

9.1 Timeline and Approach on Capacity Building Campaign and Stakeholder's Involvement

There is a direct relationship between lack of skilled manpower and system malfunction. In many cases the responsibility is assigned to the waste management staff and personnel without capacity building. The capacity building programs should include all levels of the waste management hierarchy. In this regard a guideline for training different groups should be developed at the national level. At the regional level a successful SONAGED ambassador can be chosen annually as the role model to advocate best practices and discuss effective approaches during visiting tours of authorities and different stakeholders. This mechanism can be expanded by considering target goals in different areas and annually shifting the role model within the districts to the successful implementation of the model motivating more people. In this concern the representative of the SONAGED with the best waste management practice can be made as an ambassador. Such ambassador has to be given an authority or has to go around different areas to share his experience with other ward/SONAGED representatives for fostering the overall waste management situation. Figure 9.3 gives an overview of the phase-wise action plan and the timeline for successful capacity building with stakeholder involvement.

As a **short-term goal (0–1 year)** the creation of training guidelines and training for stakeholders group for the waste management has to be prioritized. The chief training areas for different stakeholders should include the following aspects

- **Collection Staff**: Time and frequency of door to door collection, segregation of waste into fractions at source, interaction with citizens and occupational health.

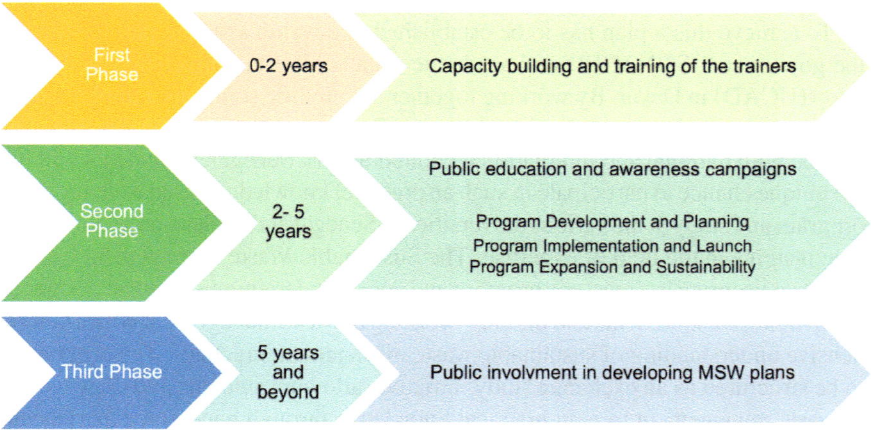

Fig. 9.3 Overview of phase-wise action plan and the timeline for successful capacity building with stakeholder involvement

- **Transportation staff**: time and frequency of transporting the waste, segregation of waste into fractions at secondary point/transfer station, interaction with collection staff and occupational health.
- **Staff at Processing Plant**: Assessing the waste received at the treatment facility including weighting and characterization, handling of fine and coarse fractions after processing, O&M of plant machinery, using personal protective equipment, Health, Safety and Environment cautiously.
- **Senior officers including sanitary inspector:** Monitoring and supervising all field activities, capacity building and welfare of staff, documentation and reporting of all activities.

At the higher hierarchy, refresher courses should be conducted for officers and supervisors at least once every 5 years. As explained earlier a SONAGED ambassador would greatly enhance the success in the management of their regional waste and can be chosen annually as the role model to advocate best practices and discuss effective approaches that a particular ambassador has chosen to obtain the success. This could be facilitated by organizing a field visit to model by the authorities/sanitary inspectors and key stakeholders from other regions.

In the midterm (2–5 years), the trained staff and workers can enormously contribute to public education and awareness-raising. To this end, waste generators have to be educated on the importance of management of waste, the need for user-fee collection, proper handling and segregation of the generated waste at source, management of organic waste at source (as animal feed, composting and biogas production), SONAGED can adopt a gradual approach to raise awareness, increase public interest in SWM issues, evaluate the community participation, make trials of different approaches, adapt the best options and maintain the approved communication schemes.

To achieve this a plan has to be established to develop a strong partnership with the government SONAGED and academic stakeholders at Universite Cheikh Anta Diop (UCAD) in Dakar. By working together and finding solutions a implementable advanced curriculum modules for a Master's Degree program can be developed and offered with national and international collaborations. Senegalese students will have the unique chance to participate in such an practical knowledge based exchange study program and study at the various universities in Senegal and in Germany gaining valuable insights in the field of expertise. The Sustainable Waste Management Capacity Building Program is a master's program and vocational training should be designed to equip Senegalese students and the crew working in waste management with a comprehensive understanding of sustainable waste management practices. The program has to be structured as an exchange study initiative, allowing students and crew working in waste management to gain practical knowledge through hands-on experiences in Senegal and Germany. Students should engage in collaborative studies at various universities, fostering a global perspective on waste management issues and solutions. The module should aim to equip students with theoretical knowledge and practical skills in sustainable waste management, foster a cross-cultural understanding by enabling them to participate in a study exchange program between Senegal and

German universities, and provide insights into strategies for waste minimization, decentralized treatment options, optimized collection, community participation, and digital technologies. The module structure should includes foundation courses on sustainable waste management, specialized courses on decentralized waste treatment options, optimized collection and transport systems, community participation and awareness, global perspectives, international case studies, exchange programs, digital technologies in waste management, integrated waste management systems, socio-economic and environmental impacts, and continuous assessment through project work, presentations, and group discussions. The program aims to empower Senegalese students with the knowledge and skills needed for effective waste management practices, contributing to the development of sustainable solutions for their communities.

The Sustainable Waste Management Capacity Building Program is a master's program and vocational training should be designed three phases: (1) Program Development and Planning, which involves establishing objectives, forming a steering committee, and developing curriculum. This includes recruiting faculty members, finalizing curriculum frameworks, and negotiating agreements with partner universities. (2) Program Implementation and Launch, which involves launching the program with the first unit of students, delivering foundation courses, facilitating study exchange programs, and implementing continuous assessment methods. This includes delivering specialized courses and organizing collaborative projects. 3) Program Expansion and Sustainability, which involves expanding program offerings, strengthening partnerships, developing internship opportunities, and conducting final evaluations. This includes graduating the first cohort, establishing alumni networks, and developing plans for program sustainability and scaling. Regular monitoring and evaluation are essential to ensure the program stays on track and meets its objectives.

As a long-term goal (5th year and beyond) community involvement plays a key role in improving solid waste management systems. The citizen should participate actively in the development of waste management schemes (Fig. 9.4). To this end the issue of new waste management methods should be brought to the public's attention (Concern) and the target neighbourhood should be invited to a community workshop (Involvement). Leaders from a variety of interest groups (regulatory authorities, residents from nearby areas, local waste management specialists and environmental/corporate representatives) should be argued to attend the workshop. During the workshop the interest groups communicate their points of agreement and disagreement to one another as well as to program designers (**Issue Resolution**). Groups should establish a list of options, which should include "no action" (**Alternatives**). Each alternative has to be explored in terms of its economic and environmental implications (**Consequences**) and the widely accepted option with more overall weightage has to be chosen (**Choice**). The measures required to carry out the program are detailed and any potential negative consequences are minimized to the extent feasible (**Implementation**). The program should be evaluated based on the community feedback regularly **(Evaluation)**.

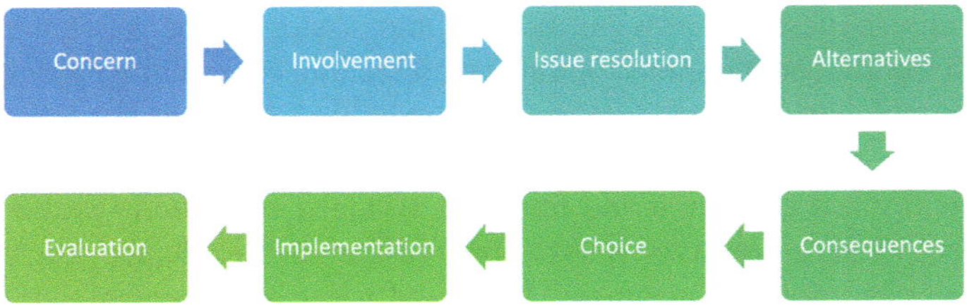

Fig. 9.4 Approach for communication campaign and public participation (during any major development of the waste management system)

Effective capacity building and active awareness-raising programs should bring into focus and put on the regions and SONAGEDs agenda. Accordingly a specific budget should be allocated for the respective activities and monitoring of such activities has to be done to evaluate their effectiveness.

It is important for academic stakeholders and research institutes to take the lead in structuring curriculums that focus solely on waste management. This will help to build capacity and ensure that training is a permanent part of the yearly curriculum. By collaborating with higher education institutions we can encourage the involvement of various stakeholders and players in the field. This will also help to build functional and practical studies for educational institutions enhancing their reach in Senegal. The University of Rostock aims to create more opportunities for cross-sector collaboration and education to enhance the environmental competence of graduates.

Chapter 10
Strategies for the Enhancement of the Funding Mechanism

10.1 Financial Sources

The development of the waste management system should be based on precise financial calculations that account for all relevant expenditures including hidden costs and profits. This critical role in the planning phase is to assure the waste management system's financial feasibility and long-term sustainability. In this concern detailed and comprehensive data on the actual costs of municipal solid waste management activities should be examined including hidden and neglected expenditures. This enables decision-makers to correctly compare current and planned services, anticipate future costs with accuracy and completely assess outsourcing possibilities. Nevertheless it is crucial to have particularly designated fund allocation for waste management and segregated financial accounting will provide a flexibility and control over the available funds.

Solid waste management activities should be planned to fulfil the legal framework. The system complexity may vary based on the local authority's financial capabilities. It is essential to ensure that all schemes are designed and implemented sustainably. Accordingly the cost recovery of projects should be guaranteed through available internal and external sources. Usually property taxes (waste taxes included) are the most reliable source of funding for SONAGED. To make solid waste management financially viable SONAGED should levy user fees on all eligible households. For the poor urban area subsidies are given so that lower user fees are levied. This should be set agreed and enforced by the steering committee comprising all stakeholders. The SONAGED should charge the services offered with appropriate tipping fees e.g. collect fees from the dumping trucks.

The sale of the final product from the recycling of organic waste, plastics, RDF, etc. is another source of funding. The quality of input material, process control, branding and marketing play important role in making revenues. Other internal resources for funding solid waste management projects are penalties. People should be informed about the offered services followed by monitoring tours. The introduction of a traffic

S. Narra et al., *Sustainable Waste Management in Senegal*, World Sustainability Series,
https://doi.org/10.1007/978-3-031-98968-1_10

light system to warn the people in a stepwise manner and raise their awareness is essential to prevent social issues.

10.2 Cost Recovery Schemes

The most challenging step in the financial viability of the waste management system is the mechanism of cost recovery. Following are the most common cost recovery schemes that could be used.

Financing from user-fee

Operational and maintenance expenditures generally require a reliable cost-recovery system for long-term sustainability. The standard approach of SONAGED for cost recovery is through standard user fee collection which is usually charged to users for the provided services. As a general rule those who generated the waste and benefited from waste management should be liable for the costs associated with its management (polluter pays principle). A charge should be levied specifically for the services made available to and used by the general public. Ideally would be to charge the entire waste service in the form of user fees to each user. It would result in a true fair distribution of the burden of collecting, treating, disposing off and following up on waste along with the prevention or restoration of the environmental damage that the waste generator could cause.

The most effective method of user fee collection is to match the willingness and the ability of users to pay the fee. In addition to willingness-to-pay, affordability-to-pay is an important marketing component for solid waste services. A failure to estimate the affordability of payments implies the danger of not being able to claim the full cost of solid waste management. Thus the use of intelligent charging mechanisms or models is essential. Based on the expenses of providing and performing a service this fee must be charged to the waste generator. User-fee should include a fixed/base fee, variable fee and EPR fee to promote a reduction in the waste generation or to make waste disposal more affordable for low-income households.

Generally the fixed costs associated with waste management are defined as those expenditures incurred depending on base waste generation and the extent to which waste disposal services are actually used. In other words, the costs are incurred primarily by providing the necessary conditions for providing the respective service. The variable costs shall be those incurred in connection with actual service provision and in particular in relation to waste volume and the range and intensity of disposal activities. In addition to base and variable fees user-fee should also incorporate certain charges that are designated as EPR (Extended Producer Responsibility) fees. EPR fee will act as an extension of the base fee and variable fee as the consumer is also responsible for the consumption of any product produced by a manufacturer.

Fixed/base fee has to be integrated to the property or asset related. Deciding the base/fixed fee has to consider various parameters viz. size of the property, type of property, i.e. private, commercial or mixed utilisation quantity of waste generated based on the size of the property etc. Variable costs come into play whenever the generator is producing a larger amount of waste than that the fixed/base fee is covering. In general fixed costs generally account for 60 to 80% of total costs whereas variable costs rarely exceed 20 to 40%.

To increase the efficiency of user fee collection, user fees may be collected directly from the customer via independent waste service bills or in conjunction with property or other utility taxes. A most practical approach for greater user-fee collection would be to introduce a digital house-tag system (similar to automatic toll collection stickers for highways on vehicles). In the house-tag system the owner of the any residential or commercial will pay a yearly user fee in advance (depending on the designated combination of base variable and EPR fee) and the difference in the paid fee and the utilised services will be corrected every year and the difference will be settled accordingly. The introduction of such a system in Senegal will ensure higher transparency and budget in advance for the smooth functioning of the waste management system.

Financing from general tax revenues

The user-fee collection efficiency could be improved if user fees are collected along with utility charges and property taxes. To achieve this it would require significant coordination between the tax authorities and the energy providers and a proper regulations and implementation flow in important stakeholders along the waste management chain as shown in Fig. 10.1. The financial sustainability of complete management of MSW services (collection, transportation, treatment and disposal) cannot be possible only by depending on user fees. Especially in the regional context of Senegal together with user-fee; funds and transfers received from state and central governments play a major role in the sustainable management of municipal waste.

In addition to government transfer and external budget support, partnership with the private sectors can also ensure the greater financial investment and sustainability of the waste management system. The private partnership could be incorporated at all stages of the waste management value chain viz., waste collection, transportation, construction and operation of waste management sites and disposal site. The private

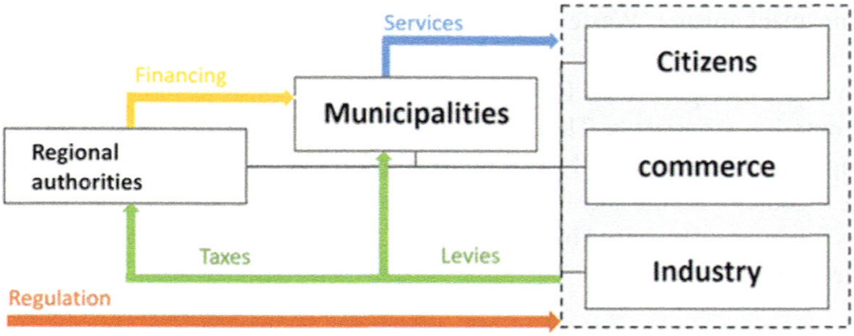

Fig. 10.1 Financing schemes from general tax revenues

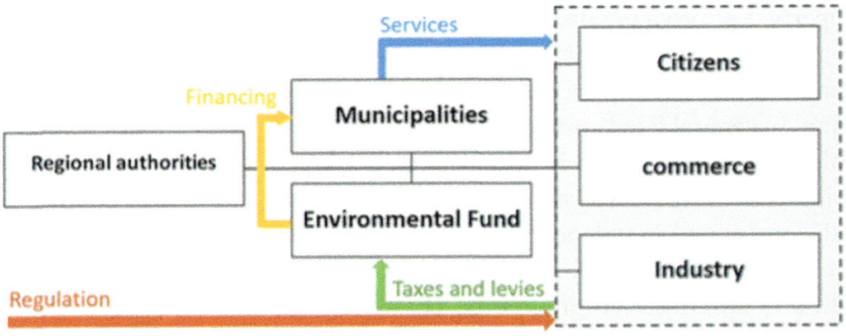

Fig. 10.2 Financial schemes from environmental funds

sector participating in the waste management can recover their cost through the service provided directly by the user and/or with the stable opportunities provided by municipalities to earn revenue from tipping fees and sales of recovered materials.

Financing from environmental funds, taxes and duties

Environmental funds or environmental taxes and duties could be an advanced mode of funding mechanism that could compliment the conventional funding mechanisms. The primary purpose of environmental taxes is to include the costs of pollutants released and the costs that are associated with the consumption of ecosystem services for the products manufactured for an economy. In general environmental pollution costs include "externalities" acquired as a result of economic activity and are not included in the direct costs incurred by producers or consumers. The collection of funds can be facilitated by charging the duties for environmental protection, penalties for environmental polluters and as additional charges for environmental authorizations of any new projects. The implementation of this strategy should follow the Senegal's hierarchy of waste manage met authorities and should included all stakeholders that have influence as seen in Fig. 10.2.

Earmarked taxes and levies

Similar to environmental taxes and duties year marked taxes and levies could be an advanced mode of funding mechanism that could complimentary the conventional funding mechanisms. Waste generation is significantly increasing because of the increasing number of tourists and pilgrims in Senegal. The problems associated with increasing waste generated by tourists and visitors can be tackled by strategically implementing a tourist tax. The collection of the funds/taxes can be facilitated directly by imposing the fees directly at the arrival of airports or could be collected by the hotels together with the stay charges. Similarly for the pilgrims visiting the religious places funds could be raised by them directly and integrate themselves into the system (suggestion).

10.3 Define Goals for Cost Recovery of MSW Services

To make MSW profitable full cost recovery is essential and for all costs for the management of organic waste should be analyzed to define realistic goals. Collection & transportation, processing & disposal, operation & maintenance costs per kg of organic waste generated from households, commercial sectors and institutional establishments should be estimated. The waste generator should be evaluated based on the amount of generated waste and other appropriate indicators. For example households based on income level, commercial and institutional sectors based on the capacity and size of the premises.

The lowest timeframe is considered for the full O&M cost recovery of the commercial and institutional sectors. For up to 3 years the collected user fees from bulk waste generators should cover 100% of the O&M cost of collection services. Similar measures could be considered for high-income level households. For middle-income level households, a longer time frame of up to 5 years is considered and for low-income level poor and slum areas, a 50% cost recovery could be set to 5 years.

10.4 Cost Recovery Schemes

The most challenging step in the financial viability of the MSW system is the mechanism of cost recovery. Three common cost recovery schemes are (1) the collection of user fees regularly e.g. monthly (2) levy collection fees annually with property tax (3) and levy charges with utility (water and electricity). The social context is an important aspect of defining the cost recovery scheme. The most effective method of user fee collection is to match the willingness and the ability of users to pay.

The standard approach of SONAGED should be to cooperate with citizens for cost recovery through periodically user fee collection (e.g. monthly collection). Fees may be fixed or variable in order to promote the reduction in waste generation or to make waste disposal more affordable for low-income households. For an instance within a single city, poor communities may be charged minimal or no fees whereas the wealthier communities, commercial operators, non-governmental organizations and institutions may be required to pay the regular user fee. The user fee in this case has to be balanced and a certain share has to be shared by the government municipalities and SONAGED.

In places where people are not willing to pay for solid waste management services, the other schemes are practical. Enforcement of the user fees should be done gradually and the SONAGED has to introduce the MSW fees as a fraction of property tax or utility charges and increase it gradually on an annual basis. A full cost recovery of the collection system in these schemes is challenging and SONAGED should try to raise the awareness of the society about the importance of a clean environment and convince them to pay the user fee periodically. It is essential to ensure that all financial schemes are designed and implemented sustainably. Sections 10.1 and

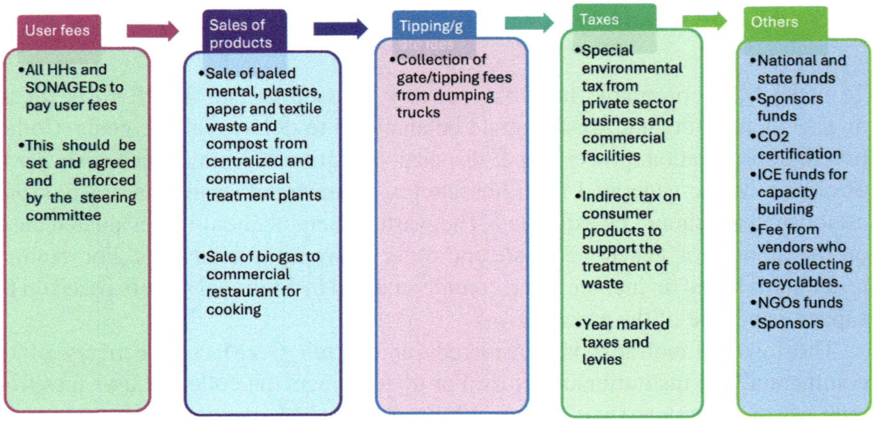

Fig. 10.3 Different sources of funding for the solid waste management system

10.2 provide the schemes to overcome most of the challenges and various types of mechanisms for the recovery of costs in the waste management system. Adaptation of such mechanisms assure the waste management system's financial feasibility and long-term sustainability in Senegal. Figure 10.3 depicts the summary of different sources for funding a solid waste management system.

Chapter 11
Impact from Pre and Post Implementation of Strategies

The Sankey diagram (Fig. 11.1) is a visual representation of Senegal's waste management situation in 2022 and the impact of integrated waste management strategies that are developed in this report by 2050. It illustrates the flow of waste from generation to management, highlighting inefficiencies and challenges.

11.1 Existing Situation of Waste Management Considering Waste Generation of Senegal in 2022

The overall municipal waste generated can be fractioned into distinct pathways for disposal and treatment. The advised fractional pathways for Senegal are organic waste, recyclable materials, refuse-derived fuel (RDF) potential and non-reusable waste. Each of these fractions requires unique approaches for effective management. These pathways are multi-divisional and the potential of every fraction can help policymakers and regional stakeholders to develop and implement an well-planed decision. It also reduces the stress on the landfills and promotes sustainable practices reducing the burden on the government.

Out of the total waste generated a major part of the organic waste can be treated further which is about 12% presently ending up in the illegal dumpsites or landfills. Similarly the fraction of recyclable waste which is 25% of the total waste generated only 50% of this fraction is currently being recycled, Fig. 11.2 give the amount of actual MSW generated and the current value waste treatment in Senegal. Focus should be placed on increasing the recycling possibilities further. Along with this the RDF from potential waste is about 15% of the total waste which can be used to recover energy and can be diverted from ending up in dump sites/landfills. Therefore the total waste that is getting landfilled can be repurposed or recycled or diverted to value addition as a sustainable waste management practice ranges from 40 to 50%

S. Narra et al., *Sustainable Waste Management in Senegal*, World Sustainability Series,
https://doi.org/10.1007/978-3-031-98968-1_11

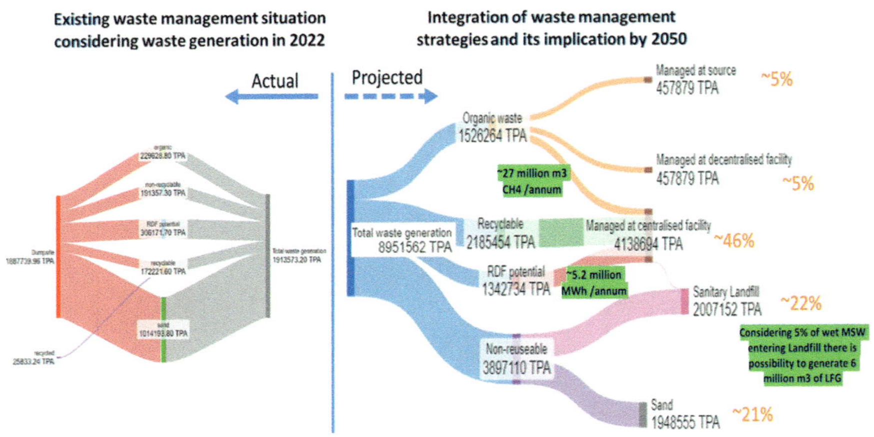

Fig. 11.1 Sankey diagram comparing the existing waste management situation in 2022 and the after implication of strategies by 2050

of the current waste. This value addition can increase in step by step manner in the near future when a functional system is setup in place.

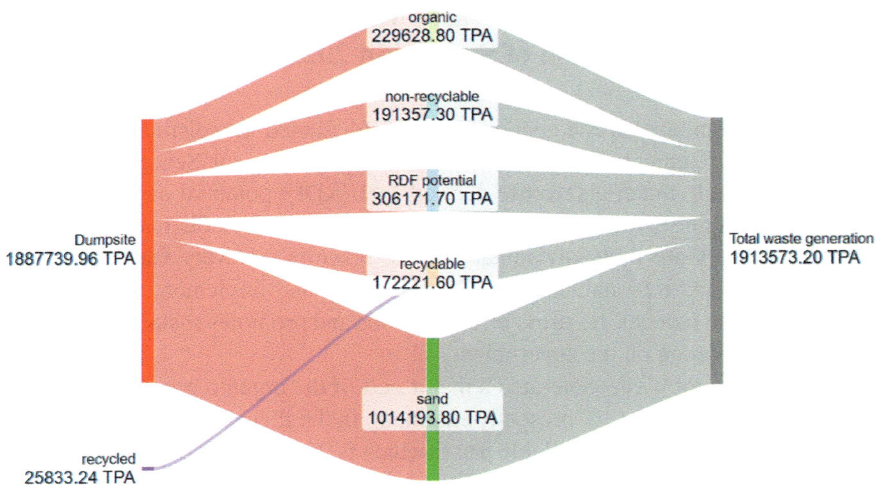

Fig. 11.2 Existing situation of waste management considering waste generation of Senegal in 2022

11.2 Integration of Waste Management Strategies in Senegal and Its Implication by 2050

The strategies recommended above can be applied to the present waste management scenario for improvement of waste fraction and its management. Since the timeline for this is variable a prediction scenario for the year 2050 is considered. For this the waste still has to be fractioned in four pathways. The total waste generation by 2050 would be at least 8,951,562 TPA, out of which 1,526,264 TPA waste can be directed into three pathways. First would be to minimize the organic waste at the source which can reduce the waste from coming to MSW chain by 457,879 TP. Other fraction of the organic waste which cannot be reduced at the source can be transported to a nearby decentralized facility for sustainable treatment and disposal. The residues of the organic treatment facilities and the contamination can be further sent to a centralized facility for further handling. By 2050 there should be enough capacity handling for a centralized facility to handle at least 5,054,452 TPA which can recover materials and energy and also recycle this amount of waste. At the end the waste that is left as a non-reusable can be sent to a sanitary landfill which properly engineered to standards. These sanitary landfills can reach up to a capacity of 3,955,707 TPA which also includes a fraction of the waste from the centralized facilities in Senegal, this is clearly demonstrated in Fig. 11.3, where by 2050 about 457,879 TPA of generated organic will be manged at source and same amount organic wastes will be treated at decentralised facilities across the country and the remaining sent to the centralised facility. Further incorporating the sand removal from the total waste and separate processing would even reduce the amount of waste entering in to landfill significantly to more than 50% less. About 80% of all generated recyclable and recovery waste will be treated at the centralised facilities when all strategies suggested are implemented and managed properly by responsible authorities.

11.3 Interlinking Effects of Adapting the Suggested Strategies

Most countries around the world are dissipating their waste by landfilling it in landfills and dumpsites without caring about its importance as a vital component of the economy. Senegal is one of those countries in which landfills accommodate 90% of the daily generated waste. At the same time, countries like Germany involve their waste in their economic sector as a regenerated wave of materials that need to be reused and harvested in different manners and techniques. Minimizing resource usage, waste cuts, and management creates the circular economy model instead of the existing linear economy (use resource–product–end of life–waste–landfill) it is good to mention here that the annual turnover in the waste sector in Germany is around 70 billion euros.

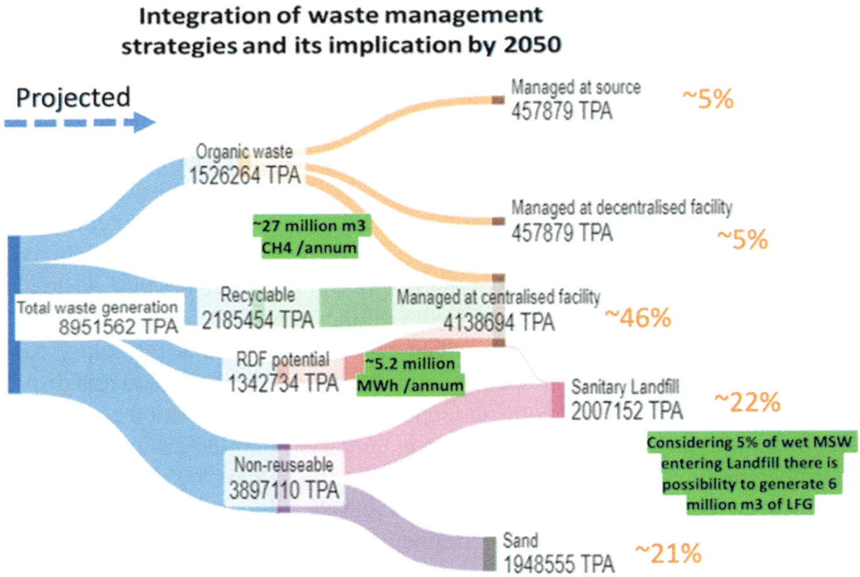

Fig. 11.3 Integration of waste management strategies in Senegal and its implication by 2050

The proposed strategy seeks to shift the existing waste sector in Senegal into a developed one by embedding all available treatment streams through two main concepts centralized and decentralized facilities.

The effect of this strategy can be divided into two terms; short-term and long-term effects.

– short-effects are interlinked among each other. It can be started as follows:

1. creation of new job opportunities for educated and non-educated manpower. Centralized facilities are quite important for this point, for example; a waste incineration plant requires around 40–50 manpower. For the strategy, 7 incinerators were proposed meaning that 350 manpower are required for the whole suggested plants. The same consideration is for the MRF facilities, where 4 facilities have been proposed that require 200 manpower. In addition, the number of drivers must be provided for the necessary trucks to transport waste from the collection points to the last destination (waste incineration, MRF, Biogas plant, landfill) around 500 drivers are required. Generally speaking, around 1000 job opportunities will be available by applying this strategy.
2. Involving the concept of centralized and decentralized facilities can increase the possibility of treating the generated organic waste. Decentralized biogas plants and compost give the potential for rural areas and the countryside to treat their own generated organic waste which in turn increases the treatment percentage of the organic waste.

3. Applying this strategy will decrease the generation rate of waste, accordingly, the spots of dump sites will also be decreased, where scientific treatment streams will be available to accommodate the daily generated waste.
4. Waste will be converted into valuable sources of materials that need to be inserted in the industrial sector and energy recovery facilities, in turn, this creates economic benefits for all institutions in charge of the waste sector.
5. For such countries as Senegal recognized with mixed waste generation manner, decentralized organic waste treatment facilities increased the opportunity to produce pure productions like compost, where the organic fraction as feedstock is directly converted into compost before it is mixed with other fractions.
6. RDF production and compost increase the revenues and the economic affordability of the waste management system. Those products can be marketed and sold in the local markets and external markets (exported) as can be seen in Fig. 11.4.

- Long-term effects:

1. CO_2 emissions have a long-term effect since the proposed facilities can be increased year by year like the decentralized biogas plants which in turn decreases the amount of CO_2 emissions. Furthermore, after 10 years of

Fig. 11.4 Interlinking effects of the proposed strategies

applying this concept, Senegalese awareness will be increased regarding waste, which lead to waste cuts and increase the circularity of waste as a source of materials and recovering energy all of which will decrease the CO_2 emissions in a long-term context. In this document, 840 kg CO_2 for each ton of MSW to be landfilled was taken, and 415 kg CO_2 for each ton of MSW for waste incineration (as an example). Each 1 million tons of waste around 400 thousand tons of CO_2 can be reduced, in other words by applying this strategy around 80% of CO_2 emissions can be hampered taking into consideration all proposed techniques.

It has been assumed that this strategy will be applied in 2025 to start with the first phase of it. As mentioned before animal feeding is the first applicable decentralized concept. The first share of the animal feeding will bw around 2% of the total daily generated waste meaning that around 60,000 ton/year of waste will be used for animal feeding, in turns this amount will save around 50,000 tons/year of CO_2, at the same time Biogas and composting will start in this phase with 1.2% around 36,000 ton/year will be used, due to that 30,000 ton/year of CO_2 emissions will be saved. The share of animal feeding and Biogas and composting as treatment stream lines will be increased to 6 and 2.5% respectively (180,000 ton/year for animal feeding and 75,000 ton/year for Biogas and compost) by 2030, in turns both techniques will contribute to save around 215,000 ton/year of CO_2 as can be seen in Table 11.1.

As has been proposed 25% of the daily generated waste will seperated by the involvement of the MRFs, meaning that around 650,000 ton/year of waste will be used for RDF production and providing recyclable materials for the industries. Using this amount of waste will save around 277,200 ton/year of CO_2 from RDF production and 242,550 ton/year of CO_2 from recycling. Thermal conversion facilities will be responsdible to recieve 40% of the waste meaning that 808,500 ton/year will be treated thermally. Note that waste combustion can save aroung 425 kg of CO_2 for each ton of waste would be combusted in comparisono with landfilling. Accordingly, by applying thermal conversion concpet with 40% as a loading rate form the waste generated around 343,612 ton/year of CO_2 can be saved.

Last but not least, in 2050 the treatment share of the organic waste will be increased to 11%, RDF production and recycling 25–40% and thermal converion 40%. Note that the projected amount of waste at that time will be around 8.9 million ton of mixed waste, due to that the total amount of CO_2 emissions that can be saved will be around 3.4 million.

As can be seen in Fig. 11.5, this strategy has a direct effect on the composition of the waste. During the first phase (the first three years) decentralized treatment approaches will start, animal feeding is the first and simplest concept that will be applied to minimize and treat organic waste at source. It is expected that after almost 2 years 2% of the yearly generated organic waste will be used for animal feeding, at the same time, Biogas as decentralized energy plants and composting will also be involved to treat around 1.2% of the organic waste. Those shares are expected to increase to 6 and 2.5% for animal feeding, Biogas, and composting respectively by 2030. It should be noted here that the projected progress depends on many parameters,

Table 11.1 Job opportunities and CO_2 emissions savings by applying the proposed strategy

Phase	Concept	Share (%)	Amount (ton/year)	CO_2 emissions (ton/year)	Job opportunities	References
2025	Animal feeding	2	60,000	50,400	100	Assumed
2030	Animal feeding	6	180,000	151,200	200	Assumed
2025	Biogas and compost	1.2	36,000	30,240	200	World Bank report (converting biomass to energy—a guide for developers and investors)
2030	Biogas and compost	2.5	75,000	63,000	200	World Bank report (converting biomass to energy—a guide for developers and investors)
2024–2030	Refuse derived fuel (RDF)	12.5	330,000	277,200	100	Assumed
2025–2030	Recycling	12.5	288,750	242,550	200	Assumed
2026–2030	Waste incineration	40	808,500	343,612.5	200	www.cleanenergywire.com
Total					1200	
2050	Animal feeding, biogas and compost	10.8	979,000	822,360	400–500	
2050	Refuse derived fuel (RDF)	12.5	979,000	822,360	200	
2050	Recycling	12.5	856,625	719,565	200	
2050	Waste incineration	40	2,398,550	1,019,348	200	

to embed the mentioned treatment concepts some regulations and laws need to be adopted by the government and presenting and facilitating financial support by law for the stakeholders, farmers, and energy suppliers who are a major component of the successfulness of this strategy at least in the decentralized part, while the centralized treatment facilities are 100% subjected to the government responsibility.

Since this strategy has taken into consideration the population growth and its consequent effect on the waste amount in 2050, due to that it has been assumed that the waste composition and the lifestyle of the Senegalean people will not change meaning that all waste fractions will occupy the same percentages without any change. In 2050 the organic fraction will still consist 12% of the whole waste composition, but the percentage (share) of the treated organic waste will furtherly increase to 10.8%.

The other portions of waste will be treated and minimized by applying the centralized treatment strategy. Multi MRF (Materials Recovery Facility) has been proposed especially in the major cities in Senegal to separate the daily generated waste into

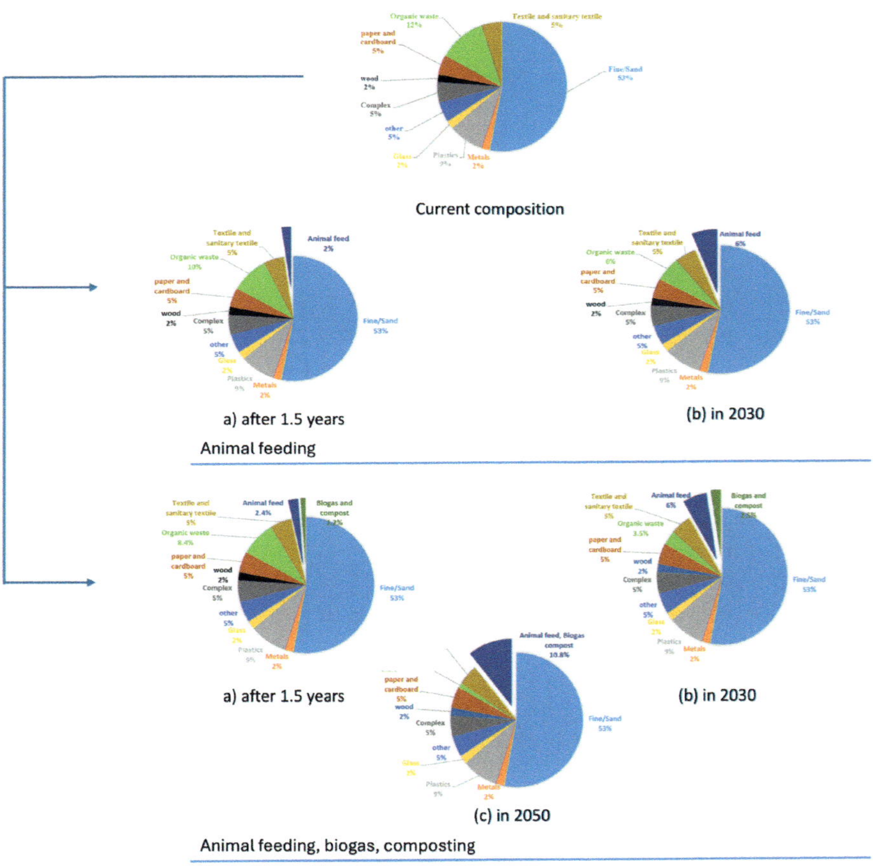

Fig. 11.5 Effect of decentralized strategy on waste composition starting with animal feeding

its aboriginal fractions. Note that most of the daily generated organic waste will be accommodated by the decentralized strategy (as mentioned previously). The waste components (fractions) will be separated into main materials by the MRFs. It has been proposed around 25% of the total waste generated will be loaded into MRFs to be separated.

Depending on the purity and MRF separation efficiency 12.5% of the separated materials can be used for RDF (refuse-derived fuel) production which consists of 2.5% Textile, 2.5% paper and cardboard, 2% wood, 2.5% complex materials (bulk materials), 2.5% others, and 4.5% plastic (see Fig. 11.6). The second 12.5% of the total waste amount can be recycled and presented as a pure materials source for industrial activities. The recyclable materials will have the same fractions' distribution of the RDF mentioned above only wood will vanish, while glass and metals will appear.

The thermal conversion processes mainly incineration, gasification, and pyrolysis will have a major role in this strategy as a major component in the centralized part and represent an efficient technique for useful energy recovery from waste worldwide. Around 40–50% of the daily generated waste will be treated in incinerators to recover its energy and produce power. Accordingly, only 15–25% of waste will be converted to landfills. As can be seen in Fig. 9.4, the waste composition will be changed sharply step by step, fine sand will have the major share of the waste composition. Waste portions are reducing due to the involvement of the treatment processes, during the period 2025–2030 the total percentage of the waste composition is around 79.25% meaning that each fraction of waste has a reduction rate of 25% due to the effect of RDF productions and recycling. Further, thermal conversion processes increased the reduction rate, due to that waste composition becoming 68.7%, where fine sand has the major fraction around 53%, while all other materials represent only 15.7%. In other words, before starting with this strategy waste composition is 100%, fine sand occupies 53% of the total composition, while the valuable materials represent 47% by involving the proposed strategy 31.3% (out of 47% see Fig. 11.7) of the valuable materials in the waste composition will be harvested and turned into useful products (energy, fuel, compost, and source of materials).In 2050 the amount of waste that can be accommodated and separated by the MRFs can be increased to 40%, and thermal conversion treatment facilities will still receive 40–50% of the waste. More amount of waste will be further treated which lead to decrease in the amount of waste that will be landfilled to 12.6% this percentage consists of 1.8% Textile, 1.8% paper and cardboard, 0.72% wood, 1.8% complex, 1.8% others, 0.72% glass, 3.24% plastic, and 0.72% metals. As a total waste composition, in 2050 the total percentage of the waste composition will be around 66% including 53% fine sand and 12.6% valuable materials.

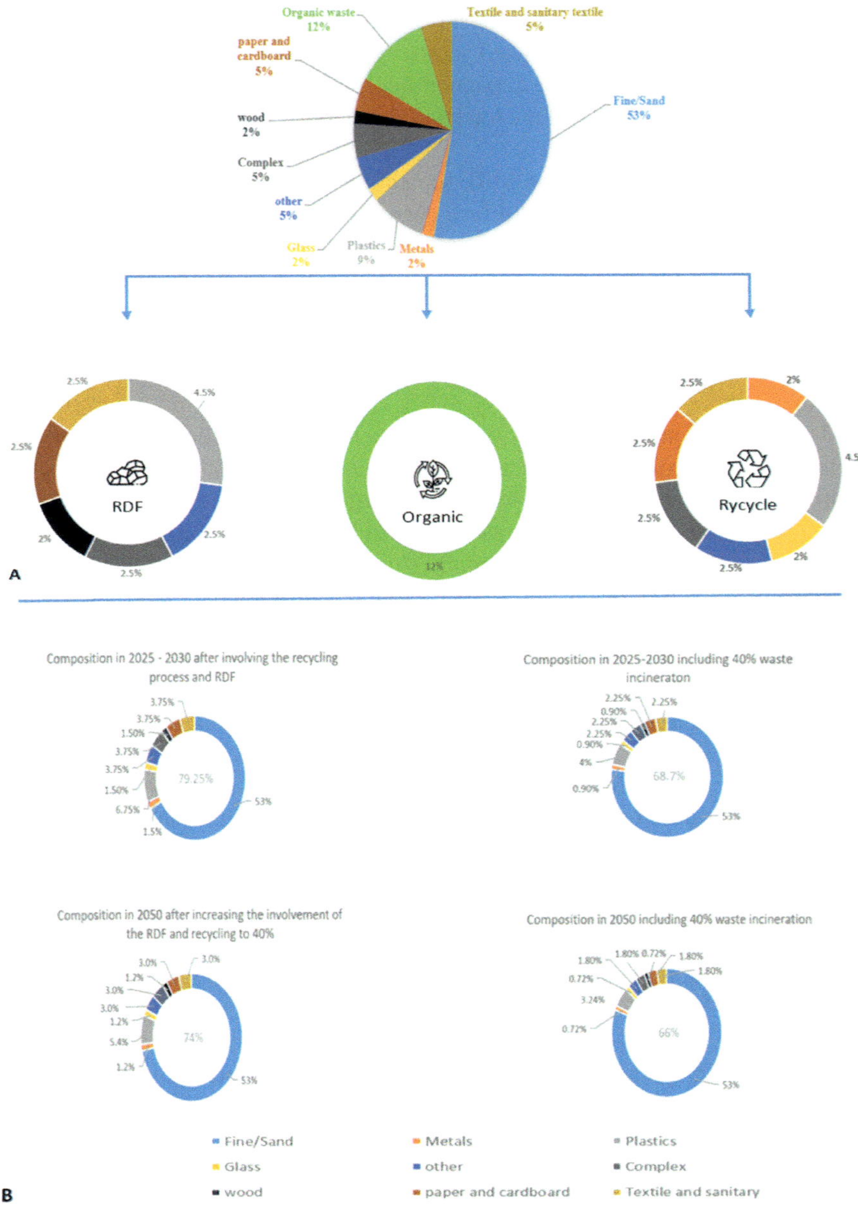

Fig. 11.6 Effect of centralized strategy on waste composition; **a** the effect of the MRFs; **b** waste composition variation due to the effect of the involvement of RDF, Recycling and thermal conversion

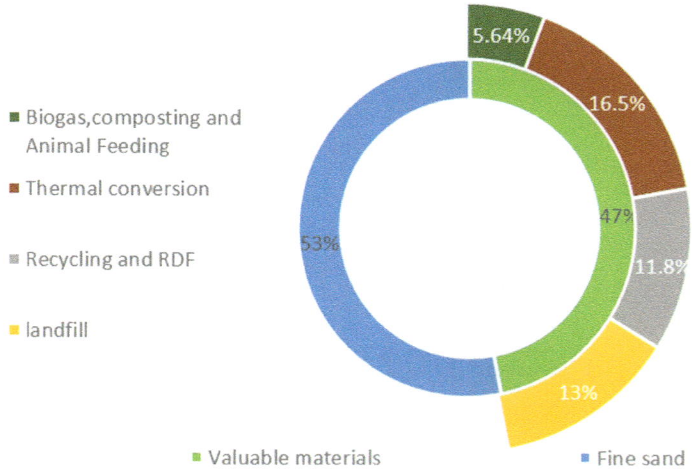

Fig. 11.7 Proposed treatment streams and their share in the valuable material

Chapter 12
Conclusion and Timelines for the Implementation of the Strategies

Senegal is located in the Sub-Saharan area which has a variety of weather conditions. Senegal is a nation located on the westernmost tip of the African continent. It has an area of 196,772 km^2. Every year the country creates more than 2.9 million tons of solid waste. Approximately 1.3 million tons (equal to 45%) of the total produced solid wastes is left uncollected. Senegal's average collection rate is 55%, which is lower than the sub-Saharan African average (65%). Senegal generates around 0.57 kg of waste per person per day (national average). The waste generating rate is higher in bigger and more industrialized cities such as Dakar where waste output averages 0.8 kg per person per day. The waste generated contains a high concentration of sand, mud, and organic material, which adds to the complexity and expense of waste treatment in the nation.

Waste collection on the other hand is variable from region to region and differs across regions in methods and coverage rates and waste collection organization depends on the size of the territory the available human material and financial means and economically. The collection and transportation services are mostly available in the departmental capitals and lack in the small municipal levels which are to be taken care by the SONAGED. Apart from SONAGED, the informal sector is also active in the SWM of the country. Overall more than 400,000 bins in the households of the urban region and more than 10,000 litter bins for the local authorities are required to be regulated.[1] As of now the waste transportation fleet is limited. The fleet has to be expanded by adding at least 700 primary waste handling vehicles 750 waste containers and 850 waste collection vehicles are required.[2] Disposal of the waste is the crucial stage where the possibility of optimizing GHG emissions, energy recovery and recovering the value of products can be analysed. In Senegal disposal of waste is primed with landfills. The burning and dumping of waste can be without proper method can be noticed throughout the country. Mbeubeuss dumping

---

[1] Sector Brief Senegal: Solid Waste Management and Recycling. GIZ, 2021.

[2] https://www.giz.de/en/downloads/SectorBrief_Senegal_Waste.pdf.

S. Narra et al., *Sustainable Waste Management in Senegal*, World Sustainability Series,
https://doi.org/10.1007/978-3-031-98968-1_12

site ranks among the top 10 dumpsites of the world. At this site approximately 1,300 tons of waste is transported by 230 trucks every day.[3]

Based on these major issues and some scientific gaps found during this study some strategic actions are suggested. Adopting these strategies would minimize the wastes to be collected at the source and reduce the stress and overload of the waste supply chain at the disposal and treatment site. This can be implemented by source segregation of waste and having clear waste streams. While having the clear waste streams the organic waste stream should be focused to recycle the nutrients back to the fields. To initiate a concrete base for the futuristic application of policies a database including all levels of information should be created and updated at a regular interval. All the Senegal need is to redefine their organizational structure to deploy a sanitary inspector. Prior to implementing any plan or treatment technology current rules must be updated to ensure proper management of the various waste stream types.

The waste management should be digitalized so that there is no data that is lost. The data should be made available to the consumer, to the waste management worker and to the authorities for the transparent exchange of the information. The steps mentioned in the strategies will not only be helpful in generating revenue and reduction in environmental stress but also they can be pinned down to generating more employment.

Efficient collection of organic waste can be ensured through a door-to-door service available after the last meal of the day (between 7 and 8 pm), followed by collection from hotels and other large waste generators between 9 and 11 pm. Additionally a daily collection service can be established to collect recyclables, hazardous and sanitary waste. The collection system is expected to cover all targeted areas in three years with partial digitalization creating job opportunities in the collection system processing facilities and dumpsites. In order for MSW to be profitable it is crucial to achieve full cost recovery. The usual method is to collect user fees periodically such as on a monthly basis. The fees can be either fixed or variable aimed at encouraging the reduction of waste generation or making waste disposal more affordable for households with low income. In places where the public is reluctant to pay for solid waste management services the implementation of user fees should be gradual with MSW fees introduced as a fraction of property tax or utility charges and increasing annually. Regular and diverse awareness-raising programs are vital to bring about behavioral change. The most effective communication strategy in this regard is face-to-face communication via collection staff and partners such as religious leaders, resident welfare associations (RWAs), community-based organizations (CBOs), non-governmental organizations (NGOs), and self-help groups (SHGs).

Establishing a centralized recycling facility to sort and bale various waste materials including sand, plastic metal, textiles and paper is an imperative step to areas sustainable waste management. By selling these materials to reputable processing industries such as SOMETA, ProPlast, Gravita Senegal Tremex, Recuver, Refemetal

[3] IDRC (2011).

and Cosem SA in Senegal would directly minimize waste to be treated and simultaneously also results in positive energy and environmental benefits of the whole operations. This includes reducing greenhouse gas emissions, enhancing soil health and safeguarding water quality. Overall this would also result as an essential and positive development for country that should not be overlooked. To ensure accountability and community engagement it is important to provide training to all stakeholders including sanitation workers and senior officials. Field visits have proven to be an effective approach in this regard. Additionally it is recommended to select a successful municipalities/regions annually as a role model (ambassadors of the country) to promote best practices and share effective approaches during visits by higher authorities from other regions/municipalities.

The strategies mentioned above can be followed by effective governance, strict policy enforcement and transparent financial flows. Conclusively the recommendations provided above hold a great potential for the overall enhancement of waste management sector and also improving public health, well-being and environmental sustainability for the country. By adopting these simple solutions Senegal could be on the road-way to a more sustainable and litter-free future.

Table 12.1 provides the summary of the developed strategies and their timeline that has been comprehensively discussed in the previous chapters of the report. These targets and action plans will guide the countries efforts to improve the existing organic waste management situation in Senegal.

Table 12.1 Developed strategies and their timeline

Time-plan						
Year	*1st year*	*2nd year*	*3rd year*	*4th year*	*5th year*	*Beyond*
Strategy: Waste minimization and waste management at the source						
Enhance the efficiency of the existing system						
Standardization and Policy						
Improve and Support animal husbandry's having potentials to recycle organic waste						
Adapting effective home-composting methods						
Capacity building and training of trainers						
Training households and monitoring						
Strategy: Improving the collection system						
Improving the system, capacity building and community involvement						
Upgrade the system and reach user fee collection goals						
100% collection coverage and different user charges						
Strategy: Closing the gap in the waste collection						
Improving the system, policy and guidelines to areas the collection of waste						
Introduce a collection scheme, introducing source separation concept and Digitalisation of money collection						
Optimizing, tracking, planning, maintenance and monitoring of the system						
Strategy: Centralized waste treatment facilities and landfills						
Existing situation analysis and optimization, and identification of location						
Setting up of controlled processing facilities (recyclables sorting unit and composting unit)						
Setting up an MRF						
Advance the technology (incineration, gasification or pyrolysis), monitor the emissions						
Maintenance and monitoring of the system						
Strategy: Transition from dumpsite to sanitary landfills						
Closing small dumpsite						
Convert hotspot dumpsite to sanitary landfills						
Close current dumping cells						
Open newly engineered sanitary cell						
Receive only stabilized organic waste in landfill						
Strategy: Digitalizing the waste management sector						
Transfer stations, collection fleet and dumpsites						
Households, bulk and commercial generators						
Treatment plants, landfills, optimization, and data accuracy						
Strategy: Stakeholder engagement, awareness-raising and capacity building						
Capacity building and training of the trainers						
Public education and awareness campaigns						
Public involvement in developing MSW plans						
Strategy: Funding mechanism						
User fees						
Collection of gate/tipping fees from dumping trucks						
Sale of products (of baled mental, plastics, paper and textile waste and compost)						
Taxation and others (state funds, ICE funds, NGO funds, CO_2 certification, etc.)						
Index for color indication						
Commencement of the task						
Achievement of the task						
Continuation beyond the achievement						

Reference

IDRC (2011) Dakar's Mbeubeuss landfill: more than meets the eye. International Development Research Centre (IDRC)

9783031989674